CASUALTIES AND CONSENSUS

*The Historical Role of
Casualties in Domestic Support for
U.S. Military Operations*

ERIC V. LARSON

RAND

The research reported here addresses the role of casualties in domestic support for U.S. military interventions. Its principal contribution is that it provides a systematic and integrated view of the major factors that are associated with public support for U.S. military operations: the operation's perceived benefits, its prospects for success, its costs, and leadership consensus or dissensus about these factors. This work should be of interest to policymakers, commanders, and planners who desire an understanding of domestic political support for the use of the U.S. Armed Forces and who are concerned about the impact of casualties on support.

The relationship between U.S. casualties and public opinion on military operations remains an important yet greatly misunderstood issue. It is now an article of faith in political and media circles that the American public will no longer accept casualties in U.S. military operations and that casualties inexorably lead to irresistible calls for the immediate withdrawal of U.S. forces. If true, this would not only call into question the credibility of the U.S. Armed Forces in deterring potential adversaries but would be profoundly important in decisions regarding the country's strategy, alliance and other commitments, force structure, doctrine, and military campaign planning.

Ten years ago, RAND published two separate reports that addressed the role of casualties in decisionmaking on U.S. military operations:

- Mark Lorell and Charles Kelley, Jr., with Deborah Hensler, *Casualties, Public Opinion, and Presidential Policy During the Vietnam War*, Santa Monica, Calif.: RAND, R-3060-AF, 1985

- Stephen T. Hosmer, *Constraints on U.S. Strategy in Third World Conflict,* Santa Monica, Calif.: RAND, R-3208-AF, 1985.

Much has happened in the intervening ten years, including the end of the Cold War, the Gulf War, and the U.S. interventions in Panama, Somalia, Haiti, and now Bosnia. This has resulted in a broader and richer set of cases to draw from in understanding the relationship between casualties and domestic support.

This report builds upon this earlier RAND work, as well as external research. It shows that there has been a great deal of continuity and consistency in the public's response to casualties in wars—including World War II and the Korean, Vietnam, and Gulf wars—and in smaller operations—including Panama and Somalia. The report presents a simple framework for understanding the correlates of domestic support and places the role of casualties in this larger context of ends, means, and leadership.

In 1994, as part of a broader study on regional deterrence, RAND published a report by Benjamin Schwarz entitled *Casualties, Public Opinion and U.S. Military Intervention: Implications for U.S. Regional Deterrence Strategies* (out of print) that addressed the relationship between casualties and public opinion. Because this report generated significant interest in the defense and foreign affairs communities, and in light of the complexity and importance of this issue, RAND believed that a more in-depth look at the question was warranted. The research reported here reveals a far more complex and subtle picture than that presented in the earlier study. Specifically, the findings that increasing casualties in Korea, Vietnam, and Somalia were associated with growing numbers of Americans favoring escalation—and rather constant low levels of support for withdrawal—were found not to be supported by the data. Although support for an orderly or gradual withdrawal often received majority support, the insight that majorities of the public tended not to prefer *immediate* withdrawal was, however, substantiated by the data. In light of this more complete analysis, RAND is replacing the 1994 report with the one presented here.

This work was sponsored by RAND corporate research funds.

Jerrold D. Green
Corporate Research Manager
International Policy Department

CONTENTS

FIGURES

TABLES

It is often said that the Vietnam War taught us that the American public is no longer willing to tolerate American casualties in U.S. wars and military operations. The public's continued willingness to accept the costs of the Second World War is occasionally compared unfavorably with Vietnam and with recent public concern about casualties in the Gulf War and Somalia, leaving aside the question of whether these cases are truly comparable.

Two contradictory corollaries are also occasionally heard. The first has it that a majority of the American public simply will no longer accept casualties under any circumstances and that the first deaths—"as the first body bags come home"—will cause a crescendo in demands for immediate withdrawal. The second has it that casualties in fact inflame the mass public, leading to an inexorable demand for "escalation to victory."

The research reported here shows that the truth is far more subtle and sensible. Majorities of the public have historically considered the potential and actual casualties in U.S. wars and military operations to be an important factor in their support, and there is nothing new in this. But the current attention to the public's unwillingness to tolerate casualties misses the larger context in which the issue has become salient: The simplest explanation consistent with the data is that support for U.S. military operations and the willingness to tolerate casualties are based upon a sensible weighing of benefits and costs that is influenced heavily by consensus (or its absence) among political leaders.

In short, when we take into account the importance of the perceived benefits and several other factors, the evidence of a recent decline in the willingness of the public to tolerate casualties appears rather thin. The Gulf War was a recent military operation in which majorities viewed important principles and interests to be at stake and showed a commensurably higher willingness to tolerate casualties than most realize. By the same token, the unwillingness of the public to tolerate very high casualties in some other recent U.S. military operations (e.g., Somalia, Haiti) has had to do with the fact that majorities—and their leaders—did not perceive the benefits or prospects to justify much loss of life.

This research set out to understand the historical role of casualties in domestic support for past U.S. wars and other military operations. Six cases that are representative of U.S. military operations in the last 55 years are analyzed: World War II; the Korean, Vietnam, and Gulf wars; Panama; and Somalia. For each case, a historical narrative was constructed describing political and military events and conditions, including U.S. casualty levels, that might have been important in shaping public attitudes toward the operation. Data were then collected and analyzed in the context of this larger narrative, including data on political and media activity and all of the contemporaneous public opinion polling that was available over the course of the operation. Other qualitative and quantitative research was also consulted wherever possible. The results are summarized in this report.

The study found that the public's aversion to loss of U.S. lives is not new, and that observed in some recent U.S. military interventions had less to do with recent declining tolerance for casualties than with the debatable merits of the military operations themselves. In fact, as Figure S.1 shows, the public has historically exhibited a highly differentiated, yet remarkably consistent, response to prospective and actual casualties in U.S. wars and military operations, including those during the most recent period:

- The public's unprecedented high apparent tolerance for casualties in the Second World War (area 1 in Figure S.1) was associated with the widely perceived gravity of the stakes that were involved, the belief that core values were being promoted, continued optimism that the Allies would defeat the Axis powers, and consistently high levels of support for the war from political leaders.

- More limited ends have justified more limited means, and the public has accordingly been somewhat less willing to accept casualties in its three limited wars in Korea, Vietnam, and the Gulf (area 2 in Figure S.1). Indeed, contrary to the conventional wisdom, recent U.S. historical experience provides a clear example of a U.S. military operation (the Gulf War) in which the interests and principles engaged were judged important enough for a majority to be willing to accept arguably high costs. This willingness was not terribly different from the public's prospective willingness to accept costs in the early days of Korea and Vietnam.

- In another recent case (Panama, area 3 in Figure S.1), majorities perceived somewhat important U.S. interests and principles at stake. Support accordingly appeared to be somewhat unaffected by the more than 20 deaths in the operation, and majorities expressed a willingness to accept somewhat higher losses if they proved necessary to ensure Noriega's capture.

- By contrast, the United States has recently undertaken—in Somalia (area 4 in Figure S.1), Haiti, and now Bosnia—the sorts of operations that have historically suffered from a low willingness to accept costs—prolonged interventions in complex political situations characterized by civil conflict, in which U.S. interests and principles are typically much less compelling or clear and success is often elusive at best. Past examples of this type include the interventions in the Dominican Republic (1965) and in Lebanon (1982–1984).

In short, it is not so much the passage of time as the prevalence of a particular class of operation that explains the apparent recent low tolerance for casualties in U.S. military operations.

THE BASES OF PUBLIC SUPPORT

The research led to a generalized model of public support that is consistent with both the historical record and the data for each of the six detailed case studies. The model is a simple one that focuses on

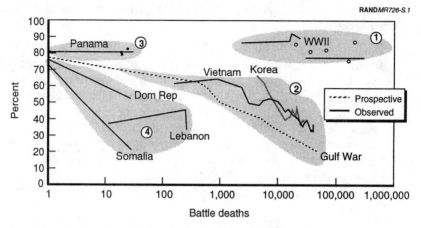

RAND*MR726-S.1*

NOTE: The wording of the questions and the data for the figure are documented in the appendix.

Figure S.1—Percent Supporting as a Function of Battle Deaths

four variables—the perceived benefits, the prospects for success (or progress), the costs, and consensus support (or its absence) from political leaders—three of which mediate the importance of the fourth (casualties) in support for U.S military operations.

At the most fundamental level, members of the public attempt to weigh ends and means in deciding whether to support military operations or asking others to sacrifice their lives. The model assists in thinking about support for an intervention by characterizing support as the result of a series of tests or questions that need to be answered collectively by political leaders and the public:

- Do the benefits seem to be great enough?

- Are the prospects for success good enough?

- Are the expected or actual costs low enough?

- Taken together, does the probable outcome seem (or seem still) to be worth the costs?

In weighing the benefits, prospects, and costs, members of the public gauge consensus or dissensus among leaders to inform their own evaluations. When leaders agree that the objectives of an operation are worth its costs and risks, this increases the likelihood of support from those who find these opinion leaders credible and trustworthy. When leaders are divided along partisan or ideological lines, however, members of the public tend also to divide along similar lines. In short, the public's views of military operations are no more homogeneous than those of political leaders, and the fault lines are reasonably predictable and consistent in both groups.

THE CHANGING LEVEL OF COMMITMENT

Both of these processes, furthermore, are dynamic: Support for a U.S. military intervention rarely remains at its initial levels and tends to fall over time (and as casualties increase). Over the course of an operation, changes in perceived benefits, prospects, casualties, and support from leaders affect the rate of declining support.

Opposition to a war or military operation can come either from members of the public who prefer a decreased commitment (de-escalation or withdrawal) or from those who believe that more should be done to achieve a successful outcome, i.e., those who desire an increased commitment (or escalation). At the extremes, some have argued that casualties and declining support have led to increasing demands for immediate withdrawal, while others (e.g., Schwarz, 1994) have argued that casualties and declining support have led to inexorable demands for escalation to victory. The data appear to contradict both extreme views, while being broadly consistent with other past RAND work and work by other scholars that demonstrates the importance of leadership and objective events and conditions in the level of the public's commitment to an ongoing military operation.

More specifically, a plurality or majority of the public during Korea and Vietnam grudgingly supported continuation of each war until a settlement and orderly withdrawal could be achieved and supported temporary escalation to break diplomatic deadlocks if the costs were not unreasonable, because they believed very important interests were at stake. In Somalia, majorities of the public in October 1993 were also unwilling to be precipitously forced out of Somalia (i.e., without recovering U.S. servicemen held hostage), but they were also unwilling to stay longer than the six months the president said it would take to accomplish an orderly withdrawal. The stakes simply did not warrant such a commitment.

When the perceived benefits are low or success is particularly elusive, the settlement that Americans prefer seems deeply, often entirely, to be bound up with the rather limited issue of getting prisoners or hostages back. Once that is accomplished, there may be little to warrant a continued presence. The case of Somalia shows, however, that, even when majorities prefer withdrawal, they may often be willing to support punitive strikes, as long as the orderly withdrawal is not hindered as a result. Such a result does not, however, suggest support for an increased commitment.

LEADERSHIP CONSENSUS AND DISSENSUS

Of course, individuals may differ greatly in their evaluations of the benefits of an operation, expectations of success and failure, and willingness to make trade-offs between benefits and costs. We might also expect them to differ in their optimism about the prospects that escalation will lead to success at low cost, which we would expect to figure in support for escalation. One reason that individuals may differ in their diagnoses and prescriptions is that they tend to turn to different leaders.

One of the most important findings of this study is the central role of leadership—and divisions among leaders—in support for military operations and preferences regarding strategy and the level of commitment. Many public opinion analyses tend to ignore leadership or to treat it simplistically as presidential manipulation of public opinion or a search to find justifications that will resonate with the public. A more sophisticated view of leadership, however, takes into account the positions of the president's opponents toward the war or military operation.

Substantial evidence supports the proposition that leadership consensus or dissensus is an essential element in the character of public support for U.S. military interventions and that leadership divisions tend to cue divisions among the public in a predictable way. In short, when there is bipartisan consensus among leaders in support of an intervention, divisions within the public are generally muted;

when there are partisan divisions among the leaders, the public
tends to become divided along the same lines.

The analysis also suggests that beliefs about benefits, prospects,
acceptable costs, and support and policy preferences can differ
across partisan or ideological groups, leading to different levels of
support and policy preferences—as one analyst pointed out, leadership and events matter:

- In the Korean War, internationalist Republicans believed Korea
  important enough to widen the war to include attacks on
  Manchuria, even at the risk of a larger war. Isolationist
  Republicans, on the other hand, did not consider Korea to be
  important enough to be worth the loss of additional lives of U.S.
  servicemen, and they were accordingly less likely to prefer escalation. Most Democrats thought the stakes important enough to
  reject withdrawal and continue the war until peace could be
  achieved but not important enough to risk a wider war. In
  Vietnam, growing polarization among leaders affected support
  and policy preferences in a somewhat similar way.

- In the buildup to the Gulf War, congressional criticism following
  President Bush's November 1990 announcement of an "offensive
  option" led to corresponding declines in public support, and the
  "rally" at the time of the Gulf War also had an important partisan
  component to it.

- In Somalia, leadership dissensus took an institutional form, pit-
  ting the Congress against the president even before the deaths in
  Mogadishu in early October 1993. Increasingly vocal congressional opposition to the operation in Somalia seems to have
  contributed in an important way to the erosion in public support. As bipartisan leadership support turned to bipartisan
  opposition and a desire for an orderly withdrawal, these preferences were mirrored in the public.

The argument presented here suggests that support—and the evaluation of benefits, prospects, and costs—is socially constructed. The
media report debates among leaders and experts to members of the
public, who consider and discuss them. The media subsequently
poll these same members of the public, informing leaders of the success of their persuasive arguments. While something of a simplifica-

tion, this characterization captures some of the most important features of how the democratic conversation between leaders and the public actually works.

IMPLICATIONS FOR POLICYMAKERS

With the end of the Cold War, the United States entered a more confusing world, and nowhere is that more apparent than in differences over the circumstances that justify the use of force. The nation lacks the clear strategic and moral purpose that garnered broad support during the Cold War and that provided a set of tests for evaluating the U.S. stakes in troubling situations. Adding to the confusion is the fact that the nation has recently intervened with force for purposes and in ways that it never has before and where arguably smaller threats, interests and principles have been engaged.

When asked to support a military operation, the American public ultimately must weigh the intangible benefits of achieving foreign policy objectives against the most tangible costs imaginable—the lives of U.S. service personnel. Leadership plays a fundamental role in this process of building and sustaining support for U.S. military interventions, although in a broader sense than it is typically conceived. Large segments of the public rely upon political leaders to vet the often complex issues involved in prospective and ongoing military interventions, and they respond predictably when the leaders they find most credible begin to question—or decide to oppose—an intervention. In short, when political and other opinion leaders fail to agree with the president that much (or any) good is likely to come of an intervention, there should be little surprise that the public also becomes divided.

The potential consequences of these recurring disagreements are quite sobering. They can lead to enduring divisions in the public and to support that is brittle and easily exploited by adversaries, thereby leading both to failed interventions and incorrect lessons for the future. Ultimately, such divisions may erode the credibility of threats of force to protect important U.S. interests. The irony, of course, is that when deterrence and coercive diplomacy fail, the costs to the nation may turn out to be even higher.

The historical record suggests that the public's tolerance for casualties and its support for U.S. wars and military operations will continue to be based upon a sensible assessment of normative and pragmatic considerations, more fully informed by leaders. When such an assessment leads to broad recognition that important national interests are engaged, important principles are being promoted, and the prospects for success are high, the majority of the American public is likely to accept costs that are commensurably high with the perceived stakes. Whenever the reasons for introducing U.S. forces lack either moral force or broadly recognized national interests, support may be very thin indeed, and even small numbers of casualties may often be sufficient to erode public support for the intervention. For in the end, most Americans do not want lives to be sacrificed for any but the most compelling and promising causes, and they look to their leaders to illuminate just how compelling and promising the causes are.

Policymakers who are mindful of the premises under which support has been given for a particular U.S. military operation will often be able to build and sustain a permissive environment for conclusion of the operation. They are also most likely to understand the constraints on—and opportunities for—presidential leadership when dramatic change occurs and initial support has eroded. But until U.S. leaders arrive at a new bipartisan consensus on the role of military force in the post–Cold War world, we should expect disagreements among leaders whenever the United States deploys its forces, and these disagreements will continue to foster divisions among the public. The absence of a larger foreign policy consensus will contribute to support that is often shallow or brittle and highly responsive to the costs in casualties. As the historical record shows, however, attributing declining support solely to casualties misses the real story.

ACKNOWLEDGMENTS

I am grateful to the following colleagues at RAND for comments on earlier drafts of this report, or the larger body of work upon which it is based: Timothy Bonds, Brent Bradley, Paul Davis, James Dewar, Bruce Don, Richard Hillestad, Stephen Hosmer, James Kahan, Irving Lachow, Thomas Lucas, Daniel McAffrey, Thomas McNaugher, John Peters, Kenneth Saunders, Peter Tiemeyer, and Kenneth Watman. I would also like to thank the following individuals outside of RAND for their comments: Richard Brody and Alexander George of Stanford University, Bruce Bueno de Mesquita of the Hoover Institution, Ole Holsti of Duke University, Louis Klarevas of the United States Institute of Peace, Harvey Sapolsky of the Massachusetts Institute of Technology, COL Richard Szafranski of Air University, and Martin Van Creveld of the Hebrew University of Jerusalem. I am greatly indebted to Mark Lorell of RAND, John Mueller of the University of Rochester, and Miroslav Nincic of the University of California, Davis for their careful reviews. I would also like to thank Phyllis Gilmore, Patrice Roberts, and Laura Zakaras of RAND for their deft editorial assistance. For data, I would like to thank Dan Amundson of the Center for Media and Public Affairs, Maura Strausberg of the Gallup Organization, Patrick Bova of the National Opinion Research Center (NORC), and George Pettinico of The Roper Center. For their invaluable assistance, I would also like to thank Susan Adler, Lorraine Gersitz, Susan McGlamery, and Roberta Shanman of the RAND library. I of course alone am responsible for any and all errors of fact or interpretation.

ABBREVIATIONS

AIPO	American Institute for Public Opinion
CACCF	Combat Area Casualties Current File
CNN	Cable News Network
Fortune	*Fortune* magazine
IISR	Institute for International Social Research,) Princeton University
JCS	Joint Chiefs of Staff
NES	National Election Studies (Survey Research Center at the University of Michigan)
NORC	National Opinion Research Center
NPR	National Public Radio
OPOR	Office of Public Opinion Research ✓
ORC	Opinion Research Corporation
POW	Prisoner of war
SRC	Survey Research Center
Time	*Time* magazine
WP, WPost	*The Washington Post*
WSJ	*Wall Street Journal*

Chapter One

INTRODUCTION

There has recently been a great deal of concern and speculation about the willingness of the American public to accept casualties in U.S. military operations. The fear is that the public has become less tolerant of casualties in military operations and has become unwilling to support operations unless they are concluded at very low cost. If true, such a conclusion would have broad implications for U.S. strategy, forces and doctrine, and for the U.S. ability to deter or coerce adversaries.

The objective of this report is to summarize data that bear on this question and to place the role of casualties into a larger framework. The report is based upon a much larger body of work that examined the roles that casualties and other factors played in influencing support for a number of U.S. military interventions. A comparative case-study approach was used to better understand the determinants of support for a number of U.S. military interventions. Detailed data, including public opinion and additional quantitative and qualitative data on political, military, and media activity, were collected and analyzed for six different wars and military actions in which U.S. ground troops were employed: the Second World War; the Korean, Vietnam, and Gulf wars; Panama; and Somalia. A less-detailed analysis of several other cases was also done to assess the robustness of the findings.

Public support for these military actions was assessed on the basis of public-opinion questions asked contemporaneously about then-ongoing or just-concluded U.S. wars or military interventions. Such data are collected by a variety of polling organizations, usually tied to media organizations. Because the many differences in wording or question structure can influence results, there is often a great deal of ambiguity in these data, which makes policy-quality public-opinion analysis a subtle, complex, and often frustrating enterprise. The current effort involved the analysis of over a thousand public-opinion questions on military operations. Although the use of public-opinion data means that we are not dealing with a great deal of precision, the interpretations reported here appear to be reasonably robust, often based upon several separate results.

Two important points are in order. The first is that individuals differ greatly in their attitudes regarding the role of force in statecraft and their willingness to ask others to sacrifice their lives to achieve what are essentially political objectives. Many of these attitudes can be traced to larger normative beliefs—beliefs that lie outside the realm of empirical data and objective analytic methods and for which there are no "right" answers. I have accordingly attempted to be respectful of the very human dimensions of this issue and equally respectful of what is ultimately a matter of conscience: how to balance relatively abstract U.S. foreign policy and security objectives against the most tangible costs imaginable—the lives of U.S. military personnel. What follows describes the patterns that seem to lie in the data and is not an endorsement of a specific political or ideological viewpoint on when casualties are or are not warranted.

The second point is that there are also profound differences in beliefs about how representative democracy works, specifically the extent to which American political leaders influence, or follow, the will of the public. Although I believe that political leaders and the public are mutually constraining, I find ample—and compelling—empirical support in the narrow issue area of U.S. military operations for the importance of opinion leadership and much weaker evidence supporting the case for a bottom-up process.

ORGANIZATION OF THIS REPORT

This study begins with a review of the six cases, focusing on the correlates of declining support. Chapter Three then examines the same cases but looks at the implications of declining support for policy preferences—support for an increased or decreased commitment ("escalation" or "de-escalation") in each intervention. Chapter Four brings leadership into the picture by describing the crucial role of leadership consensus or dissensus in the emergence of discord in the public. Chapter Five provides conclusions.

Chapter Two

THE BASES OF SUPPORT

BACKGROUND

With the end of the Cold War, the United States entered a more confusing world, and nowhere is that more apparent than in differences over the circumstances that justify the use of force. The nation lacks the clear strategic and moral purpose that garnered broad support during the Cold War and that provided a set of tests for evaluating the U.S. stakes in troubling situations. Adding to the confusion is the fact that the nation has recently intervened with force for purposes and in ways that it never before has and where arguably smaller threats, interests, and principles have been engaged. But it is in fact the confluence of a number of factors that makes support for military operations seemingly so elusive:

- When contrasted with the Soviet era, the current environment is characterized by smaller threats and stakes for the United States, and a majority of the public seems to perceive the world this way. The consequence is an apparent belief that the United States can be discriminating in its use of force and in its willingness to bear the costs of military interventions.

- While the Cold War provided a coherent framework for evaluating the importance of threats to the nation, the current environment is characterized by disagreement and conceptual confusion over the proper U.S. role and purpose in the post–Cold War world, readily apparent from the debates that erupt with each new prospective U.S. military intervention.

- The Vietnam War is now viewed by three out of four as having been a mistake, although diagnoses of what the mistakes were are no less polarized than they were at the war's conclusion.[1] Vietnam continues to cast a long shadow on prospective uses of military force and to fuel concern in some quarters about stumbling into a long, costly, and inconclusive war and, in other quarters, about entering another war that will lose the support of the American public.

- The current public opinion environment itself evidences continued support for active U.S. involvement in international affairs and opposition to isolationism—but, at the same time, the public shows a preference for concentrating on domestic economic and social issues and a desire to avoid military entanglements that will distract leaders from domestic affairs.

- In the current political environment, levels of partisanship and polarization are higher in the Congress but lower in the body politic. The result is that partisan differences among leaders may result in less predictable responses from the mass public.

- The concern about casualties among political leaders and the public, although humane, is not entirely rational—U.S. battle deaths are actually somewhat rare, typically very few, and are dwarfed by the number of deaths to U.S. service personnel from other causes.

- Although the media may not have the impact on the substantive policy preferences of the public that some impute to it, technological and other advances could have a profound effect on democratic governance. Perhaps the most important effects would be a perception among policymakers that the electronic media are shortening their decision cycles and the increasing availability of "flash" polling that often reflects little more than ephemeral and transitory opinion.

_____

[1]This was made clear by the reaction to former Defense Secretary Robert McNamara's recent book.

CASUALTIES AND SUPPORT

There is little doubt that a majority of the public is concerned about U.S. casualties when they consider support for a U.S. military intervention.[2] Consider the public opinion data in Table 2.1, which ranks concerns regarding the use of the U.S. Armed Forces and shows that the loss of American lives is the most important consideration for more than eight in ten.

It furthermore is clear that, all else being equal, prospective and observed support for a U.S. military intervention decline as expected or actual casualties increase. Figure 2.1 illustrates the typical relationship between support and casualties, in this case for a hypotheti-

Table 2.1

Importance of Various Factors in the Use of U.S. Armed Forces

No one wants our nation to get into any conflicts in the future, but as in the past, our leaders might someday decide to use our armed forces in hostilities because our interests are jeopardized. I know that this is a tough question, but if you had to make a decision about using the American military, how important would each of the following factors be to you?

	Very Important (percent)
Number of American lives that might be lost	86
Number of civilians that might be killed	79
Whether American people will support	71
Involvement by major power (e.g., USSR, PRC)	69
Length of time of fighting	61
Possibility of failure	56
Whether allies/other nations will support	56
Fact that we might break international laws or treaties	55
Cost in dollars	45

SOURCE: Americans Talk Security No. 9, October 1988.

---

[2]Categories of casualties include battle deaths, non-battle deaths, wounded in action, and prisoners of war (POWs). Unless otherwise elaborated, the word *casualties* refers to deaths due to hostile action (or battle deaths) for the remainder of this report.

cal generic U.S. military intervention.[3] It shows that, as the hypothetical costs in lives increase, fewer respondents find the number of
deaths in the intervention to be acceptable. The median respondent
(at the 50th percentile) found 100 U.S. deaths due to hostile action to
be the acceptable limit in casualties.

The figure is of questionable use to us, however: It offers no context
whatsoever in terms of the intervention's objectives or prospects for

RANDMR726-2.1

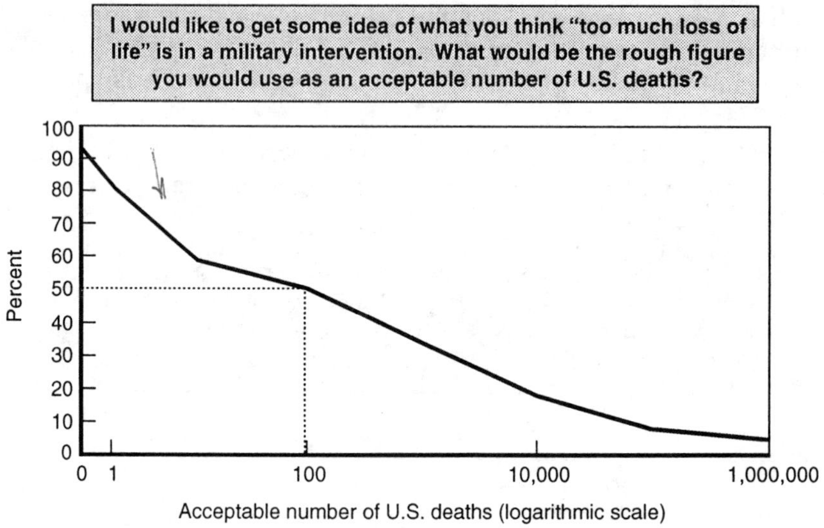

I would like to get some idea of what you think "too much loss of
life" is in a military intervention. What would be the rough figure
you would use as an acceptable number of U.S. deaths?

Acceptable number of U.S. deaths (logarithmic scale)

SOURCE: Americans Talk Issues, June 23–July 1, 1991.

NOTE: See Table A.1 in the appendix for the data from this question.

Figure 2.1—Support as a Function of U.S. Battle Deaths

---

[3]The reader should note that the figure, the first of many such figures, uses a logarithmic scale for the x-axis, which suggests more sensitivity to early casualties than
later ones. Mueller (1973) found that the logarithms of casualties in the Korean and
Vietnam wars were better predictors of support than the raw casualties because the
decline in support was steeper in the earlier part of the wars and slower toward the
end, and speculated that this was typical of support for such limited and distant wars.
As will be seen in the case studies, it appears that this intuition was well-founded; the
log of the casualties seems to explain rather well the declining support for many other
operations. See Mueller (1973), pp. 35–37 and 266.

success or of any other characteristic that describes the U.S. stakes that are engaged. In fact, there have been many U.S. military interventions, some of which were found to be unacceptable with far fewer than 100 deaths, and some of which were found to be acceptable with far more—what are the patterns in the data for actual U.S. military interventions?

Figure 2.2 plots observed support for a number of actual military conflicts—and prospective support for one other (the Gulf War)—as a function of casualties.[4]

The figure leads to two important insights. First, the rate of decline as a function of casualties varies dramatically from operation to operation. For example, judged by the two time series and the other data for the war, support hardly declined in the Second World War,

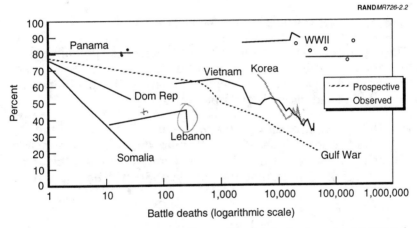

RANDMR726-2.2

NOTE: The wording of the questions and the data for the figure are documented in the appendix.

Figure 2.2—Support as a Function of U.S. Battle Deaths for the Cases Analyzed

---

[4]The reader will note that the casualty data are presented on a log scale to facilitate comparison of the cases; this understates the differences in the slopes of the low- and high-casualty operations.

while support declined rather precipitously in Somalia, losing about 30 percentage points for each increase by a factor of ten in deaths due to hostile action.

Second, the figure does not suggest a high tolerance for casualties in the past and a low tolerance in the present. For example, the curve for the Gulf War, representing prospective support given various hypothesized casualty levels, does not look terribly different from observed support as a function of casualties in Korea (1950–1953) or Vietnam (1965–1973).[5] Similarly, the intervention in Somalia (1992–1994) does not look all that different from the Dominican intervention (1965)—both showed rather steeper rates of decline as casualties grew. Lebanon (1982–1984) is an interesting case in which only a hard-core minority ever supported the operation.[6] There is clearly something at work here that bears closer examination.

A SIMPLE WEIGHING OF ENDS AND MEANS

The relationship between support and casualties can perhaps best be understood by thinking in terms of a simple model of ends and means in which leaders and members of the public determine their support on the basis of a few simple considerations:[7]

- *The perceived benefits of the intervention.* The greater the perceived stakes or interests and the more important the principles being promoted or the objectives being sought are, the higher the probability is that the intervention will be supported.[8]

---

[5]Those concerned about the comparison of data on prospective and observed support will be relieved to learn that not only were questions on prospective support as a function of casualties in the Vietnam War fairly good predictors of observed support, but the data for the Gulf War on prospective support as a function of casualties may actually have *underestimated* the public's tolerance for casualties in that war. This will be described in much greater detail later in this chapter.

[6]The slight increase in support is a temporary "rally" following President Reagan's defense of Lebanon policy in the wake of the bombing of the Marine barracks in Beirut.

[7]Milstein (1974), Jentleson (1992), Kagay (1992), Mueller (1973 and 1994), Klarevas and O'Connor (1995), and Richman (1995) reach broadly compatible findings.

[8]The clarity of the stakes, interests, or objectives can often be an important determinant of support, although this is in fact somewhat more complicated than most realize. To be sure, clarity in the objectives of a military operation has become something

Furthermore, under certain circumstances, changes in objective or mission can, in theory, either decrease *or increase* the perceived benefits.[9]

- *The prospects for success.* The higher the probability that the intervention will successfully achieve its objectives, the higher the probability is that the intervention will be supported.

- *Prospective and actual costs.* The higher the prospective and actual costs, the lower the probability is that the intervention will be supported. This is because of the increasing costs themselves and the fact that higher-than-expected costs can signal that an operation is going worse than expected.[10]

- *Changing expectations.* Initial expectations provide an anchor or frame for evaluating subsequent developments, and events that shatter this frame in a dramatic fashion (e.g., the Chinese entry into Korea, the Tet offensive of early 1968) can lead to a revision of the ends-means calculus.[11] Initial expectations—about benefits, prospects for success, costs, and support from leaders—may prove to have been unrealistic or overly sanguine. In such a case, the situation can diverge from the initial expecta-

---

of a litmus test for support from political leaders. A failure to articulate clear objectives or a disconnection between the declaratory (or original) and perceived objectives (as in Somalia) can result in high levels of criticism from political leaders and in confusion and declining support in the public. Thus, uncertainty about the stakes, interests, or objectives *may* lead to a lack of clarity about the benefits of an intervention; to the extent that this uncertainty results in a discounting or undervaluation of the expected benefits, support will usually fall. But such a lack of precision about the stakes may also *increase* support, if that ambiguity results in drawing in supporters who expect either greater or different benefits. Criticisms that President Bush should have clarified whether the United States was going to war for the oil, the nuclear weapons, or the atrocities are apposite: The apparent credibility of each justification probably contributed to the overall high levels of support by drawing support from among different groups. Had the president settled on just one justification, support would likely have been lower.

[9]In the Korean War, for example, reunification of the Korean peninsula offered more benefits than a return to the status quo ante.

[10]I agree with Mueller (1973, pp. 62–63) that casualties are a good composite indicator of the intensity and costs of an operation and that, in many cases, the public may remain unaware of the precise number of battle deaths. As will be seen, however, the public opinion data that are available often show a public that has a reasonably accurate grasp of the number of deaths that have been incurred.

[11]Michael Kagay (1992, pp. 111–112) describes the importance of expectations—in his word, *contingencies*—in the Gulf War.

tions, and cognitive dissonance, anxiety, or an urge to reevaluate the balancing of ends and means may result. When events turn out better than expected, of course, there is little reason for anxiety and much reason to applaud the outcome.[12]

- *The nature and depth of support for the intervention among the other actors.* Political leaders and members of the public are mutually influencing and constraining, and the broader and deeper the support of the other actors, the higher the probability is that an actor (e.g., member of the public, Congress) will support the intervention.[13]

In short, support can be thought of as a constant rebalancing of the benefits and prospects for success against the likely and actual costs—and a determination of whether the outcome is judged worth the costs—all informed by leaders and experts.[14] As new events occur or objective conditions change, they are interpreted by political leaders and experts, and the ends and means are reevaluated. Such a model, engendering both normative and pragmatic considerations, clearly has a great deal of intuitive appeal, but why should we believe that this is what is at work?

First, as a practical matter, U.S. military operations are typically explained and justified both in normative terms—stressing the importance of the principles and interests that are at stake—and pragmatic terms—stressing the good prospects and reasonable costs of the intervention. Such a framework captures both the "operational code" of political leaders and the enduring concerns of the public.

---

[12]Brody (1991).

[13]For example, members of the public rely extensively upon opinion leaders (the president, congressional and other leaders, and experts) to interpret and clarify events and choices and to inform their own opinions on the intervention. The president, on the other hand, gauges the attitudes of the public and Congress to determine what policies are politically feasible. Members of Congress (and the media) may gauge the receptiveness of the public to opposition arguments. The role of leadership and consensus or dissensus among leaders will be examined in Chapter Four.

[14]Documentation of other efforts to relate support for U.S. military operations to other factors can be found in Russett and Nincic (1976), Jentleson (1992), Klarevas and O'Connor (1995), and Richman (1995).

Second, a focus on the particular objectives of the operation and the perceptions of the principles and interests that are engaged establishes a connection between the objectives of a given operation and the larger purposes it is promoting.[15] A focus on principles (the mantra of the idealist school of American foreign policy) and interests (the mantra of the realists) offers a degree of simplification without unduly sacrificing analytic power. Because each school is more closely associated with a particular ideology and party (idealism with liberalism and the Democratic party, realism with conservatism and the Republican party), this focus also offers a sensible framework for examining subgroups that are likely to differ in their evaluations of a military operation.[16]

Third, such a framework is convenient for the simple reason that public opinion data are usually available. For example, a wealth of available data that describes elite and public views on foreign policy goals and vital interests evidences a high degree of stability over time in public and leadership perceptions, a high degree of consistency between opinion leaders and members of the mass public, and an observed relationship between benefits and support for military operations.[17]

What follows then describes the cases through the lens of this simple metaphor of an ends-means calculus. As will be seen, the factors

---

[15]Jentleson (1992), for example, notes typically lower levels of support for efforts to "remake" the governments of other countries than to "restrain" undesirable external aggression. This is consistent with data from the Chicago Council on Foreign Relations (CCFR) surveys that show that, typically, only about one in three views the foreign-policy goal of promoting democracy as very important.

[16]The above framework in no way suggests that all members of the public evaluate benefits, prospects, and costs—or follow leaders—in an identical fashion. Indeed, as will be seen, a great deal of evidence suggests that they do not.

[17]A simple correlation analysis of the quadrennial surveys of the 1978–1994 CCFR revealed high survey-to-survey correlations for both the public and the leaders. Nincic (1992) makes a somewhat similar observation regarding the continuity over time in these data. Similarly high correlations were found between the public and the leaders in any given survey year. While members of the public are *less inclined* than leaders to see many countries as vital interests or foreign-policy goals as being very important, both groups' overall *structures* or *rankings* of vital interests and foreign policy goals were highly correlated. See Russett and Nincic (1976) and Rielly (1995). Richman (1995) also includes benefits in his "calculus" of support; Klarevas and O'Connor (1994) find that the "justifications" offered and the expected and actual costs are important in support.

that can affect support for wars and military operations have changed little in many important respects in the last 50 years.

WORLD WAR II

Following Pearl Harbor and the U.S. declaration of war, public support for the Second World War remained high over the course of the war, as can be seen in Figure 2.3.[18]

Perceived Stakes and Benefits

In the Second World War—"the good war"—the public had an excellent cause. Of course, Japan's attack on Pearl Harbor and Germany's declaration of war on the United States contributed greatly to support for U.S. entry into the war. But support also derived from the shared perception of important stakes and vast benefits of eliminating a grave threat to U.S. security and from optimism that the outcome would be a decisive victory and punishment of the Axis powers. For many, the aims of defeating fascism and constructing international organizations to better assure peace meant the war also promoted a number of important liberal internationalist principles.[19]

Although the importance of the stakes had become clear to most of the public even before the war, it may actually have increased over

---

[18]Data on the Second World War are from Cantril and Strunk (1951), Campbell and Cain (1965), Erskine (1970), and Gallup (1972). Good analyses of public opinion data on the war can be found in Cantril (1947), Page and Shapiro (1992), and O'Neill (1993). One of the reviewers of this report suggested that comparisons of the Second World War with more recent operations must rest on a fragile foundation because of the thinness of the data on World War II, and because the war represented a transitional period in the country's international role. While both points are well taken, I have decided to leave this case study in the report, to show that even these public opinion data from an earlier era are *entirely consistent with* a model of ends and means, as just described. Page and Shapiro (1992) also examine the Second World War and the Korean and Vietnam wars and find "a rational public" responding to objective conditions and events, and leadership.

[19]Seventy-four percent had heard the phrase "United Nations" in July 1942, and 64 percent approved of the creation of a new league of nations after the war. NORC, July 1, 1942.

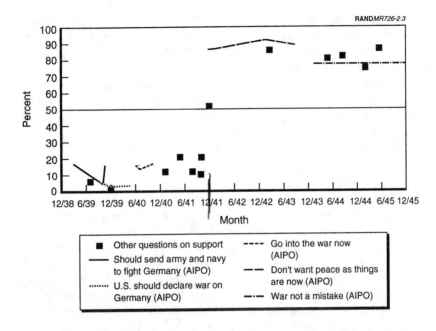

RAND*MR726-2.3*

Figure 2.3—Support for the Second World War

the course of the war. By March 1942, the Office of Public Opinion
Research (OPOR) found that 70 percent thought that, if Germany and
Japan won the war, they would "keep their armies over here to police
the United States"; by July 1942, OPOR found that 88 percent thought
this outcome likely. Eighty-seven percent thought that if Germany
won the war, it "would kill some of our business and political leaders," and 91 percent thought that "most of us would have to work for
the Nazis instead of ourselves."[20]

Further contributing to support for the war was a desire for punishment as a consequence of the Japanese sneak attack on Pearl Harbor;
such atrocities as the Bataan death march, reports of the Japanese
torture of U.S. prisoners of war, and Germany's holocaust; and the
ferocious fighting in such battles as Guadalcanal, Tarawa, Saipan,

_____
[20]OPOR, March 26, 1942, and July 15, 1942. See Cantril and Strunk (1947).

Peleliu, Iwo Jima, and Okinawa.[21] Finally, the Allies' determination to seek an unconditional surrender rather than settling for a negotiated settlement offered the sort of victory most wanted: Support for unconditional surrender ranged from about 75 to 84 percent, and majorities favored severe punishment of the malefactors.[22] Finally, for many the prospects of a postwar world organization of United Nations promised a more effective collective security system than had prevailed in the pre-war years. In short, a host of strategic and moral arguments lay behind the U.S. prosecution of the war.[23]

Prospects for Success

Also important in sustaining support, however, were the good (and improving) prospects for success. While victory may not have been in much doubt, expectations regarding the likely duration of the war were somewhat unstable and seemed to respond to events on the battlefield.[24] For example, those expecting a war of a year or less increased over the summer of 1943 (following successes in Tunisia and Sicily), fell over the winter of 1943, and then picked up again following Anzio in the spring and summer of 1944. By the fall of 1944—after the fall of Rome, the invasion of Normandy, and the liberation of Paris—fully 90 percent of those polled expected war with Germany to last a year or less.[25]

---

[21]All of these factors contributed to the demonization of the Japanese and to a belief in their treachery and savagery. Dower (1987), p. 33.

[22]NORC, February 1944; AIPO, January 31, 1945, and February 20, 1945; and *Fortune*, June 1945. Nevertheless, Mueller (1973, p. 63) reports that 20 percent in early 1944 (when Hitler still held France) responded in the affirmative when asked: "If Hitler offered peace now to all countries on the basis of not going any further but leaving matters as they now are, would you favor or oppose such a peace?" By late 1944, 88 percent felt that Japanese leaders should be punished, with torture and death the preferred punishment. AIPO, November 17–22, 1944. In mid-1945, 45 percent thought Gestapo agents and storm troopers should meet a similar fate. AIPO, May 18–23, 1945.

[23]Nevertheless, Mueller (1973, pp. 63–64) reports that only 53 percent in June 1942 felt they had a clear idea of what the war was about, although this percentage increased.

[24]In July 1942, AIPO found that 87 percent of those polled thought that the Allies would win the war, and 66 percent expected a decisive victory. AIPO, July 29, 1942.

[25]AIPO and NORC.

Expected and Actual Costs

Most expected a more costly war than the Great War had been, although in the first two years of the war, losses generally accumulated at a slow but steady rate: A little over 50,000 of the 294,000 battle deaths that the United States suffered in the war were incurred before 1944.[26] The public opinion data from the Second World War furthermore show early evidence of public concern about U.S. casualties, with most wanting more information about the toll of the war.[27]

The Allied reliance on strategic bombing was in part an attempt to avoid the massive casualties experienced during the First World War. In fact, the public seemed cross-pressured between support for an airpower-dominant strategy and a desire to put U.S. ground forces on the European continent because of doubts that airpower alone would yield success.[28] By the late spring of 1945, a majority of the public thought that a deliberative approach to concluding the war in the Pacific would result in fewer casualties and expressed a prefer-

---

[26]Forty-five percent of those AIPO polled in June 1941 expected more soldiers and sailors killed in another world war than in the first, while 31 percent expected fewer. AIPO, June 26–July 1, 1941, in Gallup (1972), p. 289. The greatest losses of the Second World War in fact occurred in the European theater in 1944, a year that included operations in Anzio, Normandy, northern France, and the Battle of the Bulge.

[27]For example, when asked how they wanted the government to handle news of U.S. losses, 73 percent of those NORC polled in December 1941 said that they wanted the government to "release news about such losses as soon as they are confirmed, so long as the news doesn't actually help the enemy." NORC, December 24, 1941. Fifty-three percent of those NORC polled in October 1943 gave at least qualified approval for publishing "even stories and pictures showing how American soldiers are suffering and dying." NORC, October 2, 1943. And 56 percent of those AIPO polled in January 1944 said newspapers and newsreels "with men dead or wounded on battlefields" should be shown. AIPO, January 6–11, 1944. Nevertheless, the October 2, 1943 NORC poll revealed that only about 10 percent said that they had actually seen pictures of G.I.s suffering.

[28]NORC's August 13, 1942 poll asked respondents "[d]uring the next two or three months, do you think the Allies should concentrate on increasing their bombing attacks on Germany, or do you think they should try to land troops somewhere in Europe to attack Germany?" Forty-one percent wanted to increase bombing, while 39 percent wanted to land troops. The rather low support for an airpower-dominant strategy may have been due to doubts that airpower alone would result in the defeat of Germany. NORC's October 6, 1942 poll found that 60 percent believed that the Allies would have to invade the continent to defeat Germany, while 28 percent thought Germany could be defeated by increasing air attacks alone.

ence for such a casualty-minimizing strategy.[29] Viewing it as a tool for quickly concluding the war without an invasion of Japan, thereby saving U.S. lives, majorities also supported the use of the new atomic bomb.[30] A majority of the public also generally was aware of the final toll of the war: In October 1945, 87 percent correctly stated that the war had been more costly than the First World War, and the median respondent AIPO polled correctly estimated that the United States had suffered between 300,000 and 500,000 deaths in the war.[31]

Conclusion

While not entirely free of domestic discord, the Second World War was "good" in very many senses of the word—it involved a bipartisan consensus about vital interests, a moral cause, and the benefits of defeating Germany and Japan—and the consequences of failing to defeat them.[32] It also evidenced continued optimism about victory. There was demonstrable concern about casualties in the war, leading to support for casualty-minimizing strategies.[33] However, other factors mediated or tempered this concern, leading to rather robust

---

[29]Forty-three percent of those *Fortune* polled in June 1945 thought that "taking more time" would result in fewer casualties than "conquering in a hurry." *Fortune*, June 1945. Seventy-nine percent of those Gallup polled in May 1945 preferred "taking time and saving lives" over "ending the war quickly despite casualties" or "ending the war quickly and saving lives." AIPO, May 15, 1945. And 58 percent said that the United States should wait until the navy and air force "had beaten them down and starved them out" before invading the main Japanese homeland. AIPO, June 27, 1945.

[30]When *Fortune* asked how they felt about the use of the atomic bomb in September 1945, 54 percent said "we should have used two bombs on cities, just as we did," while another 23 percent said "we should have quickly used many more of them before Japan had a chance to surrender." Only 5 percent said that the United States should not have used them at all, while 14 percent thought that a demonstration should have been conducted before dropping such a bomb on a city. Mueller (1973), p. 172.

[31]AIPO, October 17, 1945 and U.S. Department of Defense data. In fact, the latter was a reasonably accurate estimate—the actual number was about 407,000, including about 292,000 battle-related deaths and about 115,000 other deaths. While 38 percent said that the number of killed and wounded had been more than what they had expected when the war was started, another 42 percent said that the war had been *less* costly than had been expected.

[32]See Mueller (1973), pp. 63, 65; and Stein (1980), pp. 40–47.

[33]It also led to occasional criticism when the death toll did not seem to be justified by a particular gain in territory (e.g., Tarawa).

support in the face of the increasing toll in war dead.[34] For most, the ends and means remained in balance.[35]

THE KOREAN WAR

John Mueller's *War, Presidents and Public Opinion* (Mueller, 1973) provides the definitive analysis of public opinion data for the Korean and Vietnam wars.[36] His analysis emphasized close attention to differences in question wording and to the importance of the context and timing of public opinion polling questions.[37] Figure 2.4 presents data on trends in support for the war.[38] It shows a rally in support following Inchon in September 1950, a sharp decline following the entry of the Chinese into the war in November of that year, and a slight recovery over the spring of 1951, where it bottomed out, declining more gradually thereafter.[39] Once the front lines were restabilized at the 38th parallel and truce talks had begun, a drawn-out stalemate punctuated by occasional combat characterized the situation on the ground.

---

[34]One estimate of the decline in support as a function of casualties was about two percentage points for each increase by a factor of ten in battle deaths.

[35]Indeed, to the extent that there were disagreements among political leaders over the war, they seem mostly to have been over mobilization, treatment of labor, and other domestic aspects of the war. See O'Neill (1993).

[36]In addition to Mueller (1973), good sources of public opinion data for the war are Gallup (1972), NORC, and the Roper Center's POLL database. Good analyses of the public opinion data on Korea are found in Mueller (1973), Belknap and Campbell (1951–1952), and Page and Shapiro (1992).

[37]For example, in terms of the presidential policy at the time or of dramatic developments on the battlefield or in political circles.

[38]The labels refer to the percentage taking a "pro" position on the public opinion questions found in Table 3.1 of Mueller (1973), pp. 45–47. Question wordings for the four series are as follows. Series A (AIPO): "Do you think the United States made a mistake in going into the war in Korea, or not?" Series B (NORC): "Do you think the United States was right or wrong in sending American troops to stop the Communist invasion of South Korea?" Series C (NORC): "As things stand now, do you feel that the war in Korea has been (was) worth fighting, or not?" Series D (Minnesota poll): "Looking back over the Korean War since it started last June (in June last year, last year, two years ago, in June of 1950) would you say now that you feel the United States (we) did the right thing in sending American forces to Korea?"

[39]The increase in support in late 1952 may have been due to President Eisenhower's election and his visit to Korea in December of that year; Eisenhower had promised to bring the war to an end.

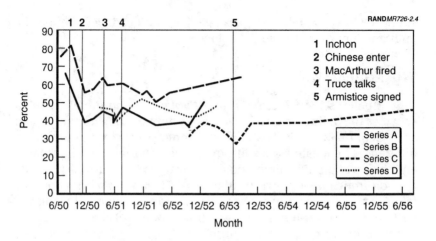

Figure 2.4—Support for the Korean War

Perceived Stakes and Benefits

Mueller (1973) argues that it was far more difficult to find convincing ideological or humanitarian reasons to justify the wars in Korea and Vietnam to the public than it had been to justify the Second World War—in short, the interests and principles engaged, or the perceived benefits of these wars, were substantially lower than those for the earlier war had been.

President Truman justified the U.S. entry into the Korean War on the basis of the importance to the United States of ensuring the security of the free world from Communist aggression and of avoiding a third world war, interests that he argued would ultimately be secured, but at potentially high cost.[40]   President Truman received initial widespread bipartisan support for his decision to enter the war.

At the onset of the Korean War, about two out of three of those AIPO polled in July 1950 saw the war as being against a dangerous com-

---

[40]In his memoirs, Truman reports that he believed that "[i]f [the North Korean invasion] was allowed to go unchallenged it would mean a third world war." Truman (1956), p. 333.

munist expansion.[41] Majorities of the public sometimes believed that a third world war against Russia and China was imminent or under way.[42] It was also an opportunity to promote the rule of law and self-determination by assisting South Korea in its own self-defense and to exercise the machinery of the United Nations in a response to the North Korean aggression.

The ambitiousness of the United Nations objectives in Korea rose and then fell. The perceived benefits from the war for much of the public probably increased with the success at Inchon, the drive to the Yalu River, and the October 1950 United Nations decision to pursue reunification of the peninsula. Following the Chinese entry into the war in November 1950, however, the United States and the United Nations returned to the war aim of restoring the border at the 38th parallel, representing a return to the more limited objective of reestablishing the border at the 38th parallel.[43]

Prospects for Success

Judging by expectations regarding the likely duration of the war, the perceived prospects for success seem to have improved dramatically following the successful defense of Pusan and the landing at Inchon.[44] Where President Truman had warned of a long and difficult struggle, many were predicting following Inchon and the drive to the Yalu in the fall of 1950 that the war would soon be concluded and that American GIs would be home by Christmas. Any hopes for such

---

[41]AIPO, July 9–14, 1950. The Minnesota poll found that of the 75 percent who approved of President Truman's action in sending American military forces to South Korea, 70 percent cited halting communism or protecting U.S. interests as their reasons. See Mueller (1973), p. 48.

[42]Gallup found that those who thought the United States was actually in World War III ranged between 45 and 57 percent in their July 30–August 4, 1950, October 1–6, 1950, and November 12–17, 1950, polls. See Gallup (1972) for the question wording.

[43]In October 1951, Republican Senator Robert A. Taft described the war as "useless."

[44]Between Gallup's July 9–14, 1950, and September 17–22, 1950, polls, the percentage expecting a war of six months or less increased by 10 percent, from 23 to 33 percent. The question asked by Gallup was: "Just your best guess, about how much longer do you think the war in Korea will last?" Unfortunately, Gallup never asked the question again. It seems likely that the perceived prospects increased further following General MacArthur's statement, before the Chinese entry into the war, that "I hope we can get the boys home by Christmas." *Congressional Quarterly* (1965).

a swift and cheap victory were soon shattered following the Chinese entry into the war, however.[45] As the situation stalemated, frustration at the lack of progress on the diplomatic front probably further eroded confidence in the outcome. In fact, by March 1951, fully 50 percent agreed that, under present conditions, neither side could win a final military victory in Korea.[46]

Costs

Most of the drop in support following the Chinese intervention corresponds to a substantial increase in the number of casualties that were taken in the bloody fighting (see Figure 2.2).[47] Furthermore, as can be seen from Table 2.2, data from February 1951 suggest a limited willingness on the part of the public to incur further battle deaths following the Chinese entry into the war: Only about one in three said they were willing to risk their own lives or those of members of their families to keep the Chinese communists from taking over Korea and other countries in Asia.

In short, even in Korea, the willingness to sacrifice American lives was bounded; limited ends justified only limited means.

Conclusion

As just described, declining support for Korea was associated with a number of factors other than combat deaths. The stakes in Korea declined from the threat of another world conflict to a limited war;

---

[45]Mueller (1973) described the impact of the Chinese intervention as having "seemed to shake from the support ranks those who were tenuous and those who felt that they could support a short war, but not a long one." Leadership opposition may also have had a role in increasing disaffection: Senator Taft said that he would prefer to quit the war in early January 1951. *Facts on File Five-Year Index, 1951–1955*, p. 308.

[46]Gallup, March 26–31, 1950.

[47]Mueller (1973), p. 52. Using multiple linear regression, Mueller found the support declined about 15 percentage points for each increase by a factor of ten in casualties (killed, hospitalized, wounded, and missing). I replicated his analysis using battle deaths rather than total casualties and found that, for every increase by a factor of ten in battle deaths, support declined between 6.79 and 8.32 points, depending on the question used.

Table 2.2

Willingness to Risk Lives in Korean War

> Would you, yourself, be willing to risk your life, or have some member of your family risk his life, to keep the Chinese communists from taking over Korea and other countries in Asia?

	Responding (percent)
Willing	34
Not willing	53
No opinion	10

SOURCE: Gallup, February 4–9, 1951.

the benefits and prospects climbed after Inchon and then fell again after the Chinese entry and over the long, drawn-out stalemate during the truce talks. As leaders turned from supporting the war, so too did members of the public.[48] In short, while costs were important, their influence cannot be untangled from these other factors.[49] While the war may well have cost him his presidency, President Truman nevertheless maintained grudging support to continue the

---

[48]For example, Senator Taft initially gave his support to Truman, but urged some congressional say in the war, and later came to criticize the president. McCulloch (1992), p. 782, 789. Belknap and Cantril (1951–1952) found high levels of partisan criticism on American foreign policy issues, including Korea, even before the Chinese entry into the war. Neustadt (1975, p. 431) provides a telling vignette from December 1950:

> In December 1950, at the wrenching turn of the Korean War, amidst Chinese attack, American retreat, renewed inflation, and fears of World War III, Truman met at the White House with the congressional leaders of both parties. Their meeting in the Cabinet room was largely symbolic, underlining events; it was an occasion for briefings, not actions. Soon after it broke up, a White House usher came to [Charles S.] Murphy's office with a memorandum found under the Cabinet table. This was a document of several pages addressed by the staff of the Senate Minority Policy Committee to Senators Taft and Wherry, the Republican leaders .... It dealt with the contingency (which had not arisen) that the President might use that meeting to seek pledges of bipartisan support for the administration's future conduct of the war. This, the memorandum argued, ought to be resisted at all costs. By Easter Recess the war could have taken such a turn that Republicans might wish to accuse Truman of treason and should be free to do so.

[49]As will be seen in Chapter Four, leadership divisions were also a critical component of declining support.

war until a negotiated settlement could be achieved, a subject that will be examined in more detail in the next chapter.

THE VIETNAM WAR

Vietnam was another major-but-limited land war in Asia—precisely the sort of war that had divided the body politic during Korea and that political leaders had forsworn after that war. It also shared with Korea the important U.S. interest of containing communist expansion and contained the risk of a dramatic increase in costs if there were to be an expansion of the war to involve China or Russia. Figure 2.5 shows the path of public support for the war.[50]

Mueller found a "rally" in support at the beginning of the war and high levels of public support into 1966. By mid-1966, however, support had declined in the wake of such events as infighting among the South Vietnamese and the emergence of vocal criticism of the war during the Fulbright hearings in early 1966. By this time, the public had also come to see that the war would not be over quickly but was instead likely to be "a long, bloody affair."[51]

In the event, U.S. forces in Southeast Asia gradually increased to over half a million. By the time of the 1968 Tet offensive, support and opposition for the war had hardened to a point at which events on the battlefield or in Washington were less likely to make an impression—support for the war was generally down to hard-core supporters and fell only perhaps an additional 10 points thereafter.[52]

---

[50]Data are from Mueller (1973), Table 3.3, pp. 54–55. Question wordings for the four series are as follows. Series A (AIPO): "In view of the developments since we entered the fighting in Vietnam, do you think the U.S. made a mistake sending troops to fight in Vietnam?" Series B (AIPO): "Some people think we should not have become involved with our military forces in Southeast Asia, while others think we should have. What is your opinion?" Series C (NES): "Do you think we did the right thing in getting into the fighting in Vietnam or should we have stayed out?" In 1964 and 1966, the question was asked only of those who said they had been paying attention to what was going on in Vietnam.

[51]Mueller (1973), p. 56. Mueller cites Harris/Newsweek poll showing that the percentage expecting a long war had increased from 54 to 72 percent between the end of 1965 and mid-1966.

[52]Mueller (1973), p. 56.

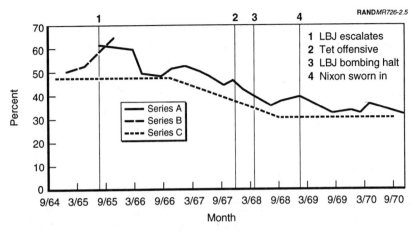

SOURCE: The data are from Mueller (1973), Table 3.3, pp. 54–55.

Figure 2.5—Support for the Vietnam War

Perceived Stakes and Benefits

The principal reasons that lay behind continued support for the war were the containment of communism, U.S. credibility, and strengthening the resolve of others to resist communism.[53] The Roper Center's February 1968 poll found that 49 percent thought one of the strongest reasons for the U.S. military effort in Vietnam was containment of communism; 33 percent cited the impact on the will of others to resist communism; and 23 percent cited the potential loss of prestige and confidence of U.S. friends and allies.[54]

---

[53]Mueller (1973) argues that initial support for the war also arose from a desire to support the president.

[54]Roper Center, February 1968, cited in Mueller (1973), p. 49. By the end of 1969, these arguments had lost none of their potency: 49 percent agreed with the statement that the war must be continued to keep communism from dominating Southeast Asia; 55 percent agreed that the United States had placed its reputation on the line and could not leave until it had ensured South Vietnam independence, and 51 percent agreed that the real reason the United States was fighting in Vietnam was because it was crucial to U.S. national security. Harris Survey No. 1961, December 1969.

Compared to the benefits of a military victory, the political objective
the Johnson and Nixon administrations pursued—achieving a negotiated settlement—probably limited the expected gains from the war
for most members of the public and may have contributed to polarization.[55] In addition to the U.S. stakes being somewhat unclear for
many, the value of the strategic stakes or consequences of a loss may
also have diminished for many leaders and members of the public
over the course of the war.[56] For others, however, the stakes may
have increased.[57]

Prospects for Success

Until the Tet offensive of early 1968, the prospects for success continued to be buoyed somewhat by administration rhetoric and

---

[55]While 29 percent of those Gallup polled in November 1965 thought the war would
end in a victory for the United States, 30 percent expected a stalemate or compromise,
and 10 percent expected a very lengthy conflict. Gallup, October 29–November 2,
1965. By May 1966, 54 percent expected the war to result in a compromise, and by
February 1968, 61 percent expected such an outcome. Gallup, May 19–24, 1966 and
February 1–6, 1968.

[56]Gallup's May 11–16, 1967 poll found the public evenly split between those who said
they knew what the United States was fighting for (48 percent) and those who said
they did not (48 percent). Seyom Brown suggests that, as the Sino-Soviet split became
more apparent and a multipolar world began to emerge and as the costs of Vietnam
rose, increasingly skeptical criticism of the strategic importance of Indochina
emerged. See Brown (1974), pp. 22–23 and 27. In fact, concern about the impact on
U.S. credibility of a precipitous withdrawal seemed to become the dominant concern
for realists. As early as 1966, such realists as George Kennan and Hans Morgenthau
were arguing that defending Vietnam to contain communism was a misapplication of
the containment doctrine. Kennan was, however, concerned about the ramifications
for U.S. credibility of a precipitous withdrawal. Mueller suggests that the benefits of
Vietnam may also have declined following the coup in late 1965 in Indonesia and
because of the inward-turning of the Chinese into their Cultural Revolution shortly
thereafter. See Mueller (1973), p. 41, and Mueller (1989), pp. 171–173 and 177–181.
According to Paul M. Kattenburg, the domestic consensus based upon containment of
the Soviet Union and communism had been fundamentally shaken by 1969. See
Lorell et al. (1985), p. 61.

[57]Nevertheless, the Roper Center's February 1968 poll found that 48 percent thought
that supporting U.S. fighting men was among the very strongest arguments for the
war, and 33 percent thought that "if the U.S. gave up, the whole expenditure of
American lives and money will have been in vain." In short, the investment of U.S.
prestige, blood, and treasure may have actually *increased* the perceived equities in the
outcome for many. This remains conjecture, since no earlier data point is available for
this question.

actions,[58] muted but growing establishment opposition, and generally favorable media coverage.[59] Tet was a military catastrophe for
the Viet Cong that effectively destroyed the organization for at least a
year.[60] Paradoxically, Tet also vitiated the rosy optimism of the fall
1967 "peace offensive" by demonstrating that the Viet Cong still had
some fight left in them.[61] While support continued to decline gradually, Tet seems to have affected the willingness to support further
escalation of the war, a subject to which I will return in the next
chapter.

Expected and Actual Costs

The perception of limited benefits from a war in Vietnam justified
only limited means for most opinion leaders and the public, as the
public opinion data in Table 2.3 suggest.

Early in the war, fewer than four in ten expressed a belief that a war
in Vietnam was worth Korea-like or higher costs. Given this limited
willingness to accept Korea-like casualties in a war in Vietnam, it is
remarkable that the Johnson and Nixon administrations were able to
continue prosecuting the war so long after Vietnam's casualty rates
had reached those of Korea. In the event, by the time of Tet, the rate
at which casualties were being incurred had mounted to the Korea-

---

[58]Nevertheless, the percentage believing that the United States was making progress
in Vietnam was less than inspiring: 31 percent in January 1966, 34 percent in July
1967, 33 percent in February 1968, and 18 percent in June 1968. Gallup January 7–12,
1966; July 13–18, 1967; February 22–27, 1968; and June 13–18, 1968.

[59]Daniel Hallin analyzed television media coverage of the war, and found that until
Tet, television coverage of the war had generally had a favorable slant. In fact, of the
statements about the Vietnam War that were presented in filmed television reports in
Hallin's sample from August 1965 to January 1968, favorable statements outnumbered
unfavorable ones by more than a two-to-one margin, and journalists' favorable editorial comments outnumbered unfavorable ones by more than a three-to-one margin.
Hallin (1986), p. 148. See also Zaller (1992, pp. 188–189, 271–273).

[60]Braestrup (1977) documents the misreporting of Tet as a military defeat for the
United States. Zaller (1992, p. 271) attributes the misreporting to journalists' surprise
at the scale of the operation in light of earlier administration statements that the Viet
Cong had been all but defeated.

[61]There is, furthermore, evidence that it was not until 1967 and 1968 that administration reports of goals and progress began to be viewed skeptically by members of the
Senate; public optimism seems to have fallen in parallel. Zellman (1971), p. 200.

Table 2.3

Prospective Casualty Tolerance in Vietnam

Asked of a sample of leaders from San Jose, Calif., in Summer 1965: At what point do you believe the U.S. objectives are not worth the cost in casualties?

	Responding (percent)
Not worth the loss or injury of a single American	32
Last year's casualty rate—100/month	10
Last year's casualty rate adjusted to increased troop strength—250/month	22
The Korean War rate—4,380/month	28
World War II rate for the U.S.—24,230/month	8

SOURCE: Voth (1965).

Asked of a national sample: Would you approve of continuing the fighting if it meant several hundred American soldiers would be killed every week?

	Responding (percent)
Approve	38
Disapprove	54
Don't know	7

SOURCE: NORC No. 876-S (February 1966).

NOTE: Percentages are of categorizable responses.

like rates that fewer than four in ten had earlier indicated a willingness to accept—an overall casualty rate in the thousands, with several hundred dying each week. It is striking that, by some measures of support, only four in ten of those polled counted themselves as supporters of the war by the time the costs reached these levels.[62]

Lorell et al. (1985) provide confirmatory evidence that the war's costs had become too high for all but a minority. After reviewing a number of other studies that associated casualties with declining support for the Vietnam War, they reported public opinion data from the Harris organization that showed that casualties, especially war dead, had increasingly become the single most troubling aspect of the Vietnam

---

[62]AIPO, March and April 1968 polls.

War.[63] By March 1969, the number of battle-related deaths had risen to over 34,000—the final toll of the Korean War—and nearly two out of three said they would have opposed the U.S. entry into the war if they had known the costs of that conflict.[64]

Conclusion

In Vietnam, the increasing costs came to be judged by majorities as being incommensurate with the expected benefits of the war and its prospects for success.[65] As in World War II and Korea, however, the role of casualties in eroding support appears to have been mediated or regulated by changing perceptions of the stakes or interests, progress in the war, and divisions among leaders.[66] Like Korea, in spite of growing misgivings on the part of leaders and the public about the growing costs of the war, presidents Johnson and Nixon were able to maintain grudging support for continuing the war until

---

[63]In July 1967, 31 percent volunteered "loss of our young men," "casualties," "loss of lives," or "killing" as the most troubling aspect of the war, and by March 1968—just after the Tet offensive—that percentage had risen to 44 percent. By comparison, only 12 percent in the July 1967 poll were most troubled that not enough progress was being made. By March 1968, 2 percent found the limitations on the war to be the most troubling aspect of the war, and 10 percent were most troubled by "too much politics," "no progress," "it's not a declared war," "why are we fighting?," or "all other responses." See also Mueller (1973), Milstein (1974), and Kernell (1978).

[64]In March 1969, Harris Survey No. 1926 asked: "If we had known the Vietnam war was going to involve the costs, the American casualties, and would last so long, would you have favored or opposed the U.S. going into the war back in 1961?" Twenty-six percent said they would have favored U.S. entry into the war, while 63 percent were opposed, and 11 percent were not sure.

[65]Lorell et al. (1985, p. 28) concluded:

> The evidence presented above shows a strong link in both the Korean and Vietnam wars between casualties and the course of public opinion regarding the war. Although there is no altogether satisfactory way to disentangle the effects of casualties from the effects of other factors with which casualties may be associated, the link is not surprising. Common sense tells us that Americans don't like to see their fathers and sons dying, especially in long wars fought over unclear or limited objectives in distant corners of the world.

[66]The slope of the decline in support as a function of battle deaths ranged from –18.6 to –19.5, depending on the question used. Although he estimated the relationship between support and total casualties (killed, hospitalized, wounded, and missing), Mueller (1973) found a comparable result—support declined by about 15 points for each increase by a factor of ten in casualties.

a negotiated settlement could be achieved, so long as the costs were minimized. This outcome will be explained in the next chapter. ✓

THE GULF WAR

There are many parallels between the Gulf War and the Korean War. The Gulf War began dramatically like Korea, with Iraq's invasion of Kuwait in August 1990.[67] The war involved important stakes, although none nearly as compelling as the containment of global communism. It also resulted in bipartisan support for the initial deployment of U.S. troops, in spite of the potential for combat. Unlike Korea, however, the Gulf War was a remarkable and swift success, achieving its objectives at costs far lower than most had expected. As can be seen in the Figure 2.6, it enjoyed high levels of support from the public.

Perceived Stakes and Benefits

Zaller (1992) has noted that the mobilization of mass support in the Gulf crisis was impressive in that it was accomplished without reference to a communist threat—the standard justification for the use of troops for the preceding 40 years—and that most people expected the war to be costly in American lives.[68]

There was, however, broad agreement that the United States had important interests in the Gulf.[69] The United States was also

---

[67]John Mueller's *Policy and Opinion in the Gulf War* (1994) provides a compilation and careful analysis of public opinion data on the Gulf War. As he did in his 1973 book, Mueller examines the impact of question wording, the options that were offered to the respondent, the timing of the poll, and other factors on public attitudes toward the Gulf War. The analysis presented here is broadly consistent with Mueller's analysis, but expands on the question of the willingness of the public to tolerate casualties. Most of the public opinion data on the Gulf War used here can be found in the appendixes of Mueller (1994).

[68]Zaller (1992), p. 269.

[69]The question CCFR asked was:

Many people believe that the United States has a vital interest in certain areas of the world and not in other areas. That is, certain countries of the world are important to the U.S. for political, economic or security reasons. I am going to read a list of countries. For each, tell me whether you feel the U.S. does or does not have a vital interest in that country . . . .

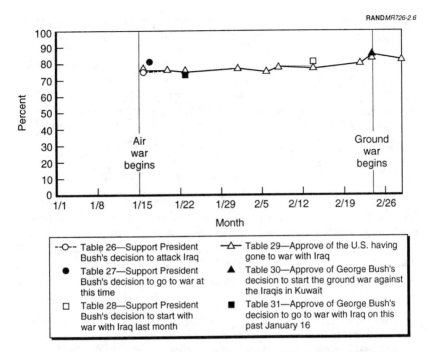

SOURCES: Black, *Los Angeles Times*, *Washington Post*, Gallup, and AP, as taken from Mueller (1994), Tables 26–31.

Figure 2.6—Approval for Starting War or Ground War, January–February 1991

arguably promoting a number of foreign policy goals or principles in the Gulf that majorities of the public generally thought were very important.[70] Majorities of the public accordingly found a number of

---

The November 1990 CCFR survey found that 83 percent of those polled believed that the United States had a vital interest in Saudi Arabia, while 77 percent believed the United States had such an interest in Kuwait. There is a high degree of consistency in these data over time: 77 percent of those polled in 1986 and 1982 indicated that the United States had a vital interest in Saudi Arabia. The equivalent percentages for the leaders were somewhat higher: 88 and 93 percent, respectively.

[70]The question CCFR asked was:

I am going to read a list of possible foreign policy goals that the United States might have. For each one please say whether you think that it should be a very

good reasons for U.S. involvement in the Gulf, as can be seen from the illustrative polling results presented in Table 2.4. √

Although there was support for a defensive deployment, support for going to war was somewhat more complex and ultimately devolved into two questions. The first was what policy to pursue to force Iraq to quit Kuwait: whether sanctions were likely to be effective at forcing an Iraqi withdrawal or whether force would ultimately be required to achieve this objective. The second was whether sanctions were likely to work and, if not, whether the situation was important enough to warrant the risks and costs of war. √

Growing support for a prospective war in the Gulf seems in part to have been the result of a declining belief that sanctions alone would achieve an Iraqi withdrawal, which precipitated both declining support for continued reliance on sanctions, and an increasing belief that war was likely.[71] That is, the leadership debate had legitimated only one alternative policy option to war—sanctions—and as the

---

important foreign policy goal of the United States, a somewhat important foreign policy goal, or not an important goal at all. How about . . . .

The fall 1990 CCFR study found the following percentages believing that each goal was very important: "defending our allies' security" (61 percent); "preventing the spread of nuclear weapons" (59 percent); "protecting weaker nations against foreign aggression" (57 percent); and "promoting and defending human rights in other countries" (58 percent).

[71]Mueller (1994, p. 58) argues that there is little evidence of a rise in support for war during the debate period:

Overall, then, Bush did not get war because he was able to swing public (or Congressional) opinion toward war—though, conceivably, he was able to arrest a deterioration of support for war.

While the evidence is not unequivocal, an examination of the tables in Mueller (1994) that contain time series for the period suggests an average six-point increase in support for war from late November to the days prior to the war. Five of the seven series examined showed an increase of more than three points over this period, while one increased by three points. Public opinion data showing a declining belief in the efficacy of sanctions is found in Mueller (1994), Tables 67, 68, and 71. Declining support for reliance on sanctions is found in Tables 72–74, 76, 78–86, 88, and 90. Increasing belief that war was likely can be seen in Tables 229–233 and 235.

Table 2.4

Percentage Saying Each Reason is a Good Reason for Getting Involved in Middle East Conflict

Please tell me if each of the following are good reasons or poor reasons for getting involved in this (Middle East) conflict.

	Agree (percent)
To deter further aggression by Iraq	78
To protect the oil supply in the Middle East	78
To force Iraq to remove its troops from Kuwait	73
To protect Saudi Arabia from Iraq	67
To remove Saddam Hussein from power	63

SOURCE: *Time*/CNN (August 23, 1990).

prospects and support for that policy declined, the public was only left with the option of war.[72]

Table 2.5 presents data on the various reasons that justified offensive action for majorities of the public—and those that did not.[73] As can be seen, the prevention of an Iraqi chemical, biological, and nuclear capability figured quite prominently in the public's willingness to use force.[74] Also believed by majorities to be good reasons for going to war, however, were restoring the Kuwaiti government, preventing an Iraqi economic threat to the United States, and preventing an Iraqi

---

[72]Relying on a cross-tabulation of data from a December 6–9, 1990 Gallup poll, Kagay (1992) found that 37 percent of Americans took a consistently hard line, favoring war over no war and force over continued sanctions, while another 30 percent took a consistently nonmilitary line over war and continued sanctions over force. This seems to have left the remaining one-third ultimately basing their judgment on the credibility of the arguments that were offered by political leaders, which would have been evaluated in part on the basis of partisan and ideological leanings. As was noted earlier, much of the rally ultimately came from the movement of those who supported sanctions to support for war. See Kagay (1992), p. 108.

[73]The result of this poll is illustrative only, but reasonably representative of the vast array of public opinion data on the subject, although other polls offered different options with different wording, which resulted in somewhat different results.

[74]The fall 1990 CCFR survey found that 59 percent considered "preventing the spread of nuclear weapons" to be a very important U.S. foreign policy goal; by 1994, the percentage was up to 82 percent.

Table 2.5

Good Reasons for Going to War, December 1990

> I'm going to read you some reasons people give for going to war against Iraq. Please tell me whether you think each is a good reason for the U.S. to go to war against Iraq or whether it is not a good reason to go to war.

	Agree (percent)
To prevent Saddam Hussein from threatening the area with chemical and biological weapons	78
To prevent Saddam Hussein from developing nuclear weapons	70
To restore the former government of Kuwait to power	60
To prevent Iraq from controlling a larger share of Mideast oil and threatening the U.S. economy	60
To prevent Iraq from ultimately attacking Israel	57
To lower oil prices	31

SOURCE: Gallup (December 6–7, 1990).

attack on Israel. By contrast, only three in ten saw going to war to lower oil prices as a good reason—few justified going to the war in crass economic terms.[75]

Support for military action was also associated with a growing litany of grievances against Iraq: taking western hostages, atrocities in Kuwait, development of nuclear and other weapons of mass destruction, and so on. For many, Saddam's self-demonizing behavior gave some credibility to President Bush's comparison of the Iraqi leader with Hitler. In fact, once the war began, the overthrow of Saddam came to be the preferred objective, an outcome that seemed all but inevitable by the end of the war.

In the event, once the air campaign began, there was a large rally in public support for the war, with nearly eight out of ten consistently

---

[75]Such a finding is not unusual—different levels of support are often offered for different justifications or when different objectives for an operation are proposed. The picture that emerges is that there are highly differentiated views in the public regarding the circumstances in which force is justified and the aims that are sufficiently important to justify the costs and risks.

supporting the war.[76] By the end of the war, few expected Saddam to remain in power, and fewer still were opposed to the war—it had been concluded more quickly and at lower cost than most had expected.

The Question of Costs

There seemed to be little doubt among members of the public that the United States would beat Iraq; the major questions were about the costs and risks of the operation and whether they were worth bearing. Few would argue that the Gulf War engaged the sorts of stakes that either the Korean or the Vietnam wars did, and fewer still would argue that the benefits of the war justified Korea- or Vietnam-like costs. With prewar predictions ranging as high as 30,000 battle deaths, the potential costs were clearly approaching Korea- or Vietnam-like numbers.[77] Was the accomplishment of the objectives seen as worth such high costs?

The willingness to accept casualties in the Gulf War was higher than most understood, but as Mueller (1994) has shown, willingness varied based upon question wording, timing, and the justifications that were offered. As in the willingness to go to war that was discussed earlier, some arguments that were offered seemed to majorities to be worth risking American lives, while others did not. Table 2.6 again shows an aversion to "blood for oil" but a willingness to accept the risk of losses to uphold the principle that countries should not get away with aggression.

---

[76]See Brody (1991, pp. 45–78) and Burbach (1995) for recent excellent treatments of the "rally 'round the flag" effect. Support for the war probably increased in part as a result of the rally effect that often follows high-visibility military actions. Panel data show that the rally was largely due to the movement of those who supported continued reliance on sanctions to active support for the war. Kagay (1992) examined panel data from the 1990–1991 National Election Studies (NES) of the Survey Research Center (SRC) at the University of Michigan and found that 44 percent of those polled had supported military action both before and after the war had begun, while 29 percent of those who wanted to rely on sanctions before the war had moved to support for the war once it had begun. Only 2 percent moved from supporting the war to preferring sanctions. NES, January 5–7, 1991 and January 17–19, 1991.

[77]A January 30–February 3, 1991, poll by ICR/Operation Real Security and the Vietnam Veterans of America Foundation found that 67 percent were aware of a Pentagon estimate that had predicted that 30,000 Americans would be killed if a ground war were started.

Table 2.6

Worth the Loss of Lives? August 1990

> There are times when it is worth the country making sacrifices in blood and money to achieve a more important return. Do you feel it is worth the loss of American lives and billions of dollars in this present (Mideast) crisis to

> —make sure American oil supplies in the Middle East are not cut off by a military power such as Iraq—or not?

	Responding (percent)
Worth it	44
Not worth it	52
Don't know	4

> —serve notice on Iraq and other aggressor nations that they cannot militarily invade and take over other nations and get away with it—or not?

	Responding (percent)
Worth it	62
Not worth it	35
Don't know	3

SOURCE: Harris (August 17–21, 1990).

There are two simple ways of understanding the relationship between casualty expectations and support, each of which leads to somewhat different insights.[78] The first is to look at questions that asked respondents whether they would support the war if it would result in a certain number of casualties.

Figure 2.7 traces the results of a number of questions in which respondents were asked whether they would support the war if certain numbers of battle deaths resulted.[79] The figure shows a decline in prospective support as the number of hypothetical casualties increases but that the rate at which prospective support declined as a

---

[78]While there is some ambiguity in these data, those presented here are representative of most of the polling data on the question.

[79]The wording of the questions can be found in Mueller (1994).

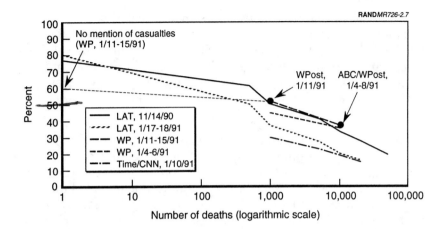

Figure 2.7—Uncertainty in the Public's Prospective Casualty Tolerance

function of casualties was roughly constant across different polling questions—although the y-intercept varies, the slopes are roughly the same.[80] Three of the prewar polls suggest that, at about 1,000 U.S. battle deaths, half of the respondents would have continued supporting the war; three polls suggest that, at 10,000 battle deaths, perhaps nearly four in ten might have remained as supporters.

The second approach to understanding the public's willingness to accept casualties is examining cross-tabulations of questions on casualty expectations and support to see whether those who supported the war expected low or high casualties; if a plurality or majority supported the war even at high prospective casualty levels, that would suggest that low casualties were not a determining factor in support. The next two tables do just this.

---

[80]The estimated slopes ranged from –9.25 to –15.0. We can of course never know whether these curves would have been good predictors of support if casualties in the Gulf War had continued to mount into the thousands or tens of thousands, but the rather consistent slopes are suggestive. The reader will recall, furthermore, that polling questions from the Vietnam War that asked about prospective willingness to tolerate Korea-like casualty rates roughly corresponded to the levels of support when Vietnam casualty rates approximated those in Korea.

Table 2.7 shows that a plurality of those polled believed that the Gulf situation was worth going to war and that, at every level of expected war dead from 3,000 to 40,000, those believing the situation was worth going to war outnumbered those who did not. Even among the 45 percent of those who were unable or unwilling to estimate the likely casualties, a plurality thought the situation was worth going to war.

Table 2.8 shows similar data collected shortly after the air war began. Eighty percent of respondents approved of the decision to go to war with Iraq, and supporters outnumbered opponents at every level of expected casualties into the tens of thousands by a margin of three to one or better. In neither case does support for the war seem to have been conditional on very low casualties.[81]

Table 2.7

Casualty Expectations by Whether Situation Was Worth War

All in all, is the current situation in the Mideast worth going to war over, or not?		
	Yes	No
Percentage of total sample	47	44

How many Americans do you think would be killed before the war was over?			
	Percentage of Total Sample	Yes (percent)	No (percent)
<1,000	6.1	45	51
≥1,000 but <3,000	6.9	44	49
≥3,000 but <5,000	2.8	64	31
≥5,000 but <10,000	4.1	64	33
≥10,000 but <15,000	4.6	51	43
≥15,000 but <20,000	2.3	55	39
≥20,000 but <30,000	6.5	48	44
≥30,000 but <40,000	3.7	51	42
≥40,000 but <50,000	2.8	40	53
≥50,000	15.4	41	53
Don't know or refused to answer	44.9	46	41

SOURCE: Gallup (January 3–6, 1991).

---

[81]This appears to be a robust finding, also turning up in other polls.

Table 2.8

Casualty Expectations by Approval of Decision to Go to War

Do you approve or disapprove of the United States' decision to go to war with Iraq in order to drive the Iraqis out of Kuwait?		
	Approve	Disapprove
Percentage of total sample	80	15

Now that the U.S. has taken military action against Iraq, do you think that the number of Americans killed and injured will be...		
	Percentage of Total Sample	Agree (percent)	Disagree (percent)
Less than 100	11.7	86	5
Several hundred	24.3	90	7
Up to a thousand	16.0	92	7
Several thousand	28.7	86	12
Tens of thousands	4.7	70	21
Don't know or refused to answer	14.7	79	12

SOURCE: Gallup (January 17–20) 1991).

In spite of this apparent willingness to accept more than what most would consider to be "low" casualties, majorities of the public supported a host of efforts to minimize casualties. For example, majorities supported the use of diplomacy, economic sanctions (until their success came into doubt), the use of force if U.S. hostages were being killed, and a prolonged air war.[82]

---

[82]Harris/NPR's December 10–13, 1990 poll found that 61 percent believed that "a diplomatic settlement providing Saddam Hussein some face-saving way to get out of Kuwait, such as giving him a small part of Kuwait with access to the Persian Gulf" would be an "honorable way to avoid American casualties." When the question was asked again in their January 10–14, 1991 poll, 53 percent thought it would be an honorable conclusion. Nevertheless, when Harris/NPR's February 21–24, 1991 poll showed that 75 percent preferred a ground invasion to get rid of Saddam Hussein "even if that involved heavy U.S. casualties" over the Russian proposal, in which Iraq would have left Kuwait but Saddam would have remained in power. While 85 percent of those *Time*/CNN polled on August 23, 1990 favored military action if Iraq started killing hostages, only 54 percent favored military action to release the hostages. Harris/NPR's February 8–10, 1991 poll found that 87 percent thought that "heavy bombings of Kuwait and Iraq will be capable of weakening the Iraqi ground forces so that your casualties in a ground invasion will be much lower" and 74 percent were willing to risk a longer war if it would result in lower casualties. Nevertheless, Gallup's

Conclusion

The success of the war and the U.S. efforts to minimize casualties were rewarded with high levels of support, and most found the costs that had been incurred to have been worth what was accomplished.[83] As suggested by the earlier discussions of the importance of the prospects for success and the data on prospective support as a function of casualties, support for the war would almost certainly have declined had the war gone much worse than expected. But as long as President Bush was willing—like presidents Truman and Johnson before him—to accept the potentially high political costs of continued prosecution of the war, he probably also would have maintained grudging support for an orderly conclusion to the war and would have avoided calls for a precipitous withdrawal. Like Korea and Vietnam, however, it seems likely that the price of this grudging support would have been the minimization of further U.S. casualties until a negotiated settlement could be achieved.

PANAMA

Panama represents a somewhat smaller operation than the Gulf War, but the U.S. objectives there were also successfully achieved rather quickly and at low cost. Other cases that were similar in this respect include Lebanon (1958), the Mayaguez (1975), and Grenada (1983).[84]

Perceived Stakes and Benefits

The objectives in Panama addressed a number of short- and longer-term problems that had bedeviled both the Reagan and Bush administrations: ensuring the safety of American citizens; dealing a blow to Panamanian drug trafficking; restoring the viability of Panamanian

---

January 17–20, 1991 poll found that 80 percent thought that air forces alone would not win the war and U.S. and allied ground troops would be needed.

[83]Looking back on the war, 77 percent of those *Time*/CNN polled in March 1991 said the war had been worth the costs; by May, a comparable question from NBC/*Wall Street Journal* found 74 percent believing the war had been worth it.

[84]The U.S. intervention in Lebanon in 1958 was generally to reassure the Lebanese government during a period of turmoil in the Middle East. The intervention in Grenada aimed at stabilizing the political situation to better assure the security of Americans there, and ejecting Cuban engineers who were building an airfield there.

democracy; and ensuring the continued security of the Panama Canal.[85] The benefits of the operation ultimately became personalized, however, and the criterion for success for most was the overthrow and capture of Noriega himself—like Saddam Hussein, Noriega was rather accomplished at self-demonization.

With the harassment and death of U.S. citizens in December 1989 and what appeared to many to be a Panamanian declaration of a state of war between Panama and the United States, President Bush was able to make a compelling argument that Americans were in danger in Panama,[86] a justification for the use of force that typically receives high levels of support.[87] Majorities of the public also believed, however, that unless the overthrow and capture of Noriega also resulted, the intervention was unlikely to be worth its costs: 58 percent said that U.S. actions in Panama could not be considered successful unless Noriega was captured, but 72 percent were optimistic that Noriega would eventually be caught.[88] Most of the military objectives of the operation were in fact achieved within a day, and combat operations were concluded rather quickly—the public

---

[85]Eighty percent of those Gallup/Newsweek polled in December 1992 said that attacks on Americans were good grounds for sending U.S. troops to another country. Roper's February 1994 poll found 76 percent favoring the use of U.S. armed forces "to strike back when Americans in a foreign country are attacked." See also Richman (1995) on this point. Ninety percent of those in a June–July 1989 Gallup poll believed that "making it harder for illegal drugs to get into this country" was a "very important" idea as a way to halt the drug epidemic in the United States. In their September 7–18, 1988, poll, Americans Talk Security found that 71 percent thought that a commitment of U.S. forces to the Panama Canal was worth the potential cost in money and in lives. Nevertheless, only 47 percent of the respondents to the Los Angeles Times' December 21, 1989, poll thought that the U.S. action was vital to the U.S. national defense.

[86]According to an ABC News poll of December 20, 1989, 65 percent thought that Americans in Panama were in a great deal or a fair amount of danger, while another 28 percent thought that they were in some danger, and 68 percent thought that the shooting of a U.S. soldier by Panamanian forces signaled increasing danger to Americans in Panama. An ABC News poll the next day found that 87 percent thought that the reasons President Bush had given for invading Panama were good enough.

[87]Roper Starch Worldwide asked the following question in its February 12–26, 1994 poll: "There are numerous situations—some strictly military, some other than military—in which our armed forces could be used. As I read each situation, please tell me whether you definitely support using our armed forces, probably would, probably would not, or definitely would not ... To strike back when Americans in a foreign country are attacked." Forty-six percent said they would definitely favor using the armed forces, while 30 percent said they would probably favor such action.

[88]ABC News, December 21, 1989.

was quickly given tangible evidence of success.[89] It was not until January, however, that General Noriega was finally taken into custody.

Costs

The bulk of U.S. combat deaths—21 of the 23 deaths that were incurred—were incurred in the first day of combat. Judging from the data in Figure 2.8, if casualties had mounted and Noriega had *not* been captured, majorities might not have felt the operation to have been worth its costs in U.S. lives. As it was, of course, Noriega was captured without substantially higher casualties, and rather than the 55–65 percent support the intervention might have received if Noriega had been captured with higher losses, Panama was consistently supported by eight out of ten.

Conclusion

The invasion of Panama ultimately enjoyed very high levels of support, of course, because it achieved quickly and at low cost objectives that were considered to be reasonably important by most political leaders and a majority of the public.

SOMALIA

Somalia was an intervention that promised vast humanitarian benefits and high prospects for success at little or no cost in U.S. lives and, accordingly, benefited from bipartisan congressional support. It was also an intervention in which U.S. combat and other forces were engaged for over a year.[90]

---

[89]Sixty-four percent of those ABC News polled on December 21, 1989 thought that the action had been more of a success than a failure. The relatively low percentage is probably due to the fact that many were withholding judgment until it was clear whether Noriega would be captured.

[90]Other cases where U.S. ground combat forces were engaged for a few months or longer include the Dominican Republic (1965) and Lebanon (1982–1984). U.S. forces were in the Dominican Republic from late April 1965 to late September 1966 and in Lebanon from August 1982 to February 1984. See OASD (FM&P) (1993), p. E-1, and Clodfelter (1992, pp. 1075–1077).

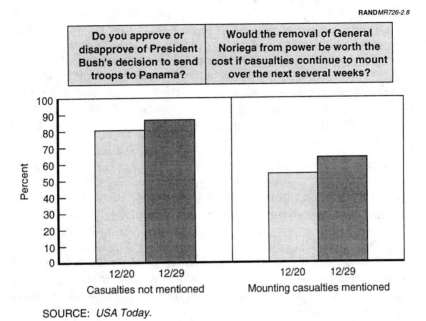

RAND*MR726-2.8*

| Do you approve or disapprove of President Bush's decision to send troops to Panama? | Would the removal of General Noriega from power be worth the cost if casualties continue to mount over the next several weeks? |

SOURCE: *USA Today.*

Figure 2.8—Support for Panama With and Without Casualties Mentioned

Perceived Benefits and Prospects

Although very few perceived a vital interest in Somalia, three out of four initially supported the operation because of the vast humanitarian benefits of saving hundreds of thousands of Somali lives.[91] In fact, until the disintegration of the security situation over the summer and early fall of 1993, the operation generally lived up to expectations, and bipartisan support—or permissiveness—held.

---

[91]CCFR surveys have found that combating world hunger is typically viewed as a very important foreign policy goal by majorities of those polled: 63 percent in 1986 and 56 percent in 1994.

The initial objectives had essentially been accomplished by the time of the transition to United Nations control in May 1993, probably yielding most of the benefits that were expected from the operation. At that time, a broader mission of supporting political reconciliation among warring factions—dubbed "nation building"—was pursued, and Aidid resisted the disarmament of his clan, evidently in the belief that he was being isolated from the political reconciliation process. The result was increasing violence over the summer, beginning with an attack in June 1993 that killed more than twenty Pakistani peace-keepers. By late June 1993, public support had fallen to about 50 percent.[92]

The mission subsequently seemed to shift to what came to be called "warlord hunting"—attempts to capture the warlord Aidid—and U.S. Army Rangers undertook a series of unsuccessful raids. Between late June and September, then, support seems to have fallen from about 50 to about 40 percent. With the failure either to maintain a stable environment in Mogadishu or to seize Aidid, the prospects for success had fallen. By September, only 36 percent of those polled thought that the U.S. efforts in Somalia were "under control."[93] By October 5, only 25 percent thought that the U.S. operation in Somalia had been successful.[94]

Costs

The four additional U.S. deaths in August and three more at the end of September meant the cost in deaths due to hostile action had nearly tripled in the space of less than two months.[95] With the 18 deaths in Mogadishu in early October, the costs had more than doubled again, resulting in high levels of congressional and media criticism and further declines in public support.

---

[92]In their June 21–24, 1993 poll, CBS/New York Times found that 51 percent approved of the president's handling of Somalia.

[93]By comparison, 52 percent thought the United States was too deeply involved. NBC News/Wall Street Journal, September 10–13, 1993.

[94]Gallup/CNN/USA Today, October 5, 1993.

[95]There were four deaths due to hostile action through March 1993, but between August and September the toll had climbed to eleven.

Figure 2.2 showed that declining support for Somalia was associated with cumulative casualties, but this figure masked the important role of changing objectives and increasingly elusive prospects for success.[96] In short, the incremental benefits of the operation appear to have declined for most; the prospects had also declined; and the costs had risen above those that most had initially expected or been willing to support.

Somalia and the Myth of the "CNN Effect"

The conventional wisdom has it that media reporting on Somalia drove both foreign policy decisionmaking and public opinion. To better gauge the relationship between media reporting and presidential decisionmaking, I performed a quantitative analysis of media reporting on Somalia and examined the sequencing between presidential decision announcements regarding Somalia and increases in media reporting levels. The analysis suggested that the frequently heard argument that "the CNN effect"—i.e., that high levels of media reporting on the human misery in Somalia prior to the presidential decisions drove presidential decisionmaking—does not appear to be supported by the data.

The White House announced its decision to begin emergency airlift of famine relief to Somalia on August 14 (see Figure 2.9).[97] In all of July 1992, there were a total of only three news reports on Somalia on ABC, CBS, and NBC news combined and only one report on CNN. During the two-week period prior to the announcement (August 1–14), ABC, CBS, and NBC together carried a total of only ten reports— or about three apiece on average; CNN carried a total of only nine reports in their round-the-clock programming for the period. In short, there does not appear to be a large increase in reporting prior to the airlift decision. In fact, the figure shows the greatest increases in media reporting levels after the White House announcement.[98]

---

[96]It also neglects the increasingly vocal bipartisan congressional opposition to the intervention, a subject that will be discussed in Chapter Four.

[97]"Statement by Press Secretary Fitzwater on Additional Humanitarian Aid for Somalia, August 14, 1992."

[98]Indeed, in a conference held at George Washington University in the spring of 1995, Andrew Natsios, formerly of the United States Agency for International Development

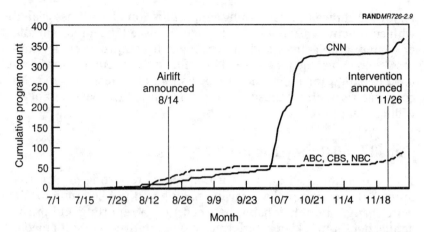

SOURCES: LEXIS/NEXIS; Center for Media and Public Affairs. The author is grateful to Dan Amundson of the Center for Media and Public Affairs for data on ABC, CBS, and NBC news reporting.

Figure 2.9—Cumulative Television Reporting on Somalia

Regarding the period prior to the November 26 announcement that U.S. troops would be deployed to Somalia, the figure shows a dramatic increase in CNN reporting on Somalia (but not in commercial television reporting) in October 1992, the period when clan fighting prevented United Nations efforts to deliver relief. But this increased reporting on CNN had tapered off by about the third week of October and did not pick up again for more than a month, until after the November 26 announcement. In short, in neither case do the data confirm the conventional wisdom of a CNN effect; media reporting levels increased after, not before, the presidential decisions on Somalia.[99]

(USAID), noted that USAID was unsuccessful in its efforts to draw media attention to Somalia in the spring of 1992, well before the issue became salient to the media.

[99]There was, however, an increase in reporting levels in October 1992.

Conclusion

The story of Somalia can perhaps best be understood by recognizing that the premises upon which support had been built—near-certain accomplishment of a limited humanitarian objective at low to no cost—were eroded by subsequent events, and this loss of support was compounded by a failure on the part of U.S. leaders to understand and attend to the eroding bases of support. The consequence was a situation in which few believed that what might be accomplished was worth additional losses—the benefits were never perceived by most to have warranted much loss of life.[100] In short, Somalia provides us with another case in which other factors were important in affecting the importance of casualties in declining support. As will be seen in the next chapter, however, the data do not support the view that majorities of the public wished to withdraw immediately, but neither do they support the contention that the public favored an increased or escalated commitment. The evidence suggests a far more subtle set of attitudes.

Some Other Comparisons

There are two other cases in which the public's tolerance for casualties might be productively compared to the experience in Somalia: the U.S. interventions in the Dominican Republic (1965) and in Lebanon (1982–1984).[101]

The Dominican Republic (1965). Over the course of the intervention in the Dominican Republic in 1965, support fell from around 75 percent to between 36 and 52 percent, depending on question wording. As with the other cases, a number of factors other than casualties

---

[100]While about four in ten in October 1993 consistently said that, "[g]iven the loss of American life, the financial costs, and other risks involved . . . sending U.S. troops to make sure food got through to the people of Somalia was worth the cost" (CBS News, October 6, 1993, October 6–7, 1993, October 18–19, 1993), 60 percent of those *Time*/CNN polled on October 7, 1993 agreed with the statement that "Nothing the U.S. could accomplish in Somalia is worth the death of even one more soldier."

[101]At the time research for this report was completed, Haiti was still an ongoing operation, and the United States had just begun deploying ground troops to Bosnia. Preliminary public opinion data suggest that both of these cases are also in the class of small, low-benefit operations: neither is perceived by a majority to engage U.S. vital interests or moral obligations, and less-than-majority support for each has resulted.

also contributed to the declining support. Most importantly perhaps, the U.S. objective soon changed from rescuing Americans to intervening against leftist forces in the civil war, which resulted in high levels of congressional criticism of the Johnson administration, especially during the hearings chaired by Senator Fulbright.[102] In short, while the data points in the figure show a rather strong decline in support as a function of casualties, the paucity of the data and the greater importance of leadership criticism suggest that the role of casualties was modest in comparison to other factors.[103]

Lebanon (1982–1984). The U.S. objective in Lebanon changed over the course of the operation: the rescue of the Palestine Liberation Organization, assistance in an Israeli withdrawal, and, ultimately, support for the beleaguered Gemayel government while keeping the Syrians from consolidating a grip on Lebanon. Page and Shapiro (1992, pp. 259–260) report that 57 percent of the public in September 1982 told Gallup/*Newsweek* they approved of President Reagan's decision to send U.S. Marines to Beirut to "help keep the peace" and to "encourage a withdrawal of Israeli, Syrian and PLO forces." The public opinion data suggest that, following the U.S. evacuation of the Palestine Liberation Organization, Lebanon was never supported by a majority of the public—typically only about 40 percent were supporters.[104] In spite of this low level of public support, the Reagan administration was able to continue the operation largely on the basis of conditional support from the Congress.[105] While there is

---

[102]The hearings Senator Fulbright held on the Dominican intervention in September 1965 were highly critical of the intervention.

[103]The slope of –78.8 in Figure 2.2 suggests that support declines by about 79 points for each increase by a factor of ten in the number of deaths due to hostile action.

[104]The appearance in Figure 2.2 that support for Lebanon was insensitive to casualties is an artifact of there not being much room left for support to fall when U.S. deaths climbed; from the very beginning, support appears to have come only from hard-core supporters, and even before the bombing of the Marine Barracks, public support for Lebanon was a partisan affair. While only 37 percent of those Gallup polled October 7–10, 1983, felt that the United States had not made a mistake in sending the Marines to Lebanon, this percentage was 53 percent among Republicans, 29 percent among Democrats, and 36 percent among independents. The slight increase in support for Lebanon is associated with a very modest "rally" following President Reagan's speech and congressional acquiescence to a continued presence following the Marine barracks bombing.

[105]In a sequence of events that the Somalia intervention paralleled ten years later, the Congress had taken action to limit the operation in Lebanon in the month prior to the

some evidence that the leaders and the public thought that the United States had a vital interest in Lebanon, there is less evidence that they thought the objectives being promoted in Lebanon were either very important or likely to be achieved.[106] Following the deaths in the bombing of the Marine barracks, support rallied slightly and then declined.

CHAPTER CONCLUSIONS

The conventional wisdom of a recent decline in the willingness of the American public to accept casualties is inadequate. There is nothing new in this concern: A majority of the public have historically considered the potential and actual casualties in U.S. wars and military operations to be an important factor in their support. Less well understood, however, is the fact that the importance of casualties to support has varied greatly across operations; when important interests and principles have been at stake, the public has been willing to tolerate rather high casualties.

In short, when we take into account the importance of the perceived benefits, the evidence of a recent decline in the willingness of the public to tolerate casualties appears rather thin. The Gulf War was a recent military operation where majorities viewed important principles and interests to be at stake and showed a commensurably higher willingness to tolerate casualties than most realize. By the same token, the unwillingness of the public to tolerate very high casualties in some other recent U.S. military operations (e.g., Somalia, Haiti) has had to do with the fact that majorities—and their

---

Marine barracks bombing in October 1983. In the case of Lebanon, the Congress gave the Reagan administration 18 months to finish the operation, and that agreement held until the spring of 1984.

[106]In CCFR's October 1982 survey, 55 percent of respondents from the public and 74 percent of the leaders said that they thought the United States had a vital interest in Lebanon, while 36 percent of the public sample and 46 percent of the leaders said they thought the United States had a vital interest in Syria. Only 34 percent of the public thought that "protecting weaker nations against foreign aggression" should be a very important foreign policy goal. Twenty-six percent thought "helping to bring a democratic form of government to other nations" should be a very important goal, and 43 percent said "promoting and defending human rights in other countries" should be a very important goal.

leaders—did not perceive the benefits or prospects to justify much
loss of life.

The public's aversion to losses of U.S. life in recent U.S. military
interventions thus has less to do with a recent decline in the public's
willingness to accept casualties than the debatable (and debated)
merits of the cases themselves. In fact, the public shows a highly
differentiated view of recent U.S. military operations that argues
against the simplistic view that the public is unwilling to accept
casualties under any circumstances:

- The recent U.S. historical experience provides a clear example of
  a U.S. military operation (the Gulf War) in which the interests
  and principles engaged were judged important enough for a
  majority to be willing to accept rather high costs, and this willingness was not terribly different from the public's prospective
  willingness to accept costs in the early days of Korea and
  Vietnam.

- In another recent case (Panama), majorities perceived important
  U.S. interests and principles at stake, and a majority accordingly
  were willing to accept greater losses if they proved necessary to
  capture Noriega.

- By contrast, the United States has recently undertaken (in
  Somalia, Haiti, and now Bosnia) precisely the sort of operations
  that have historically suffered from a low willingness to accept
  costs—prolonged interventions in complex political situations in
  failed states characterized by civil conflict, in which U.S. interests
  and principles are typically much less compelling, or clear, and
  in which success is often elusive at best. Past examples of this
  type include interventions in the Dominican Republic (1965) and
  in Lebanon (1982–1984).

This chapter has also presented evidence showing that support for
U.S. wars and military operations is dynamic and subject to a number of factors in addition to U.S. casualties. Specifically, it suggests
that the perceived benefits and prospects may often be just as—or
more—important than casualties in determining support and that
these factors affect the importance of casualties in eroding support.
There is strong evidence that declining perceived benefits or

prospects erode public support. In short, Americans do not want to sacrifice lives for causes they do not consider compelling.

In the next chapter, I will turn to the implications of falling support for policy preferences, and in Chapter Four, I will discuss the key role of leadership consensus or dissensus in both support and policy preferences.

Chapter Three

POLARIZATION OVER COMMITMENT

It seems clear that support typically declines as casualties rise, but that other factors—perceptions of benefits, broadly conceived, and prospects for success—can affect the importance of casualties in this decline, leading in different cases to higher or lower responsiveness to the growing toll. This insight says nothing, however, about the changing level of commitment of the public as support declines. Opposition to a war or military operation can come either from members of the public who prefer a decreased commitment (de-escalation or withdrawal) or from those who believe that more should be done to achieve a successful outcome, i.e., those who desire an increased commitment (or escalation).

Two bits of conflicting conventional wisdom are occasionally heard on this subject. The first, the more commonly expressed view in the national security community, has it that, as casualties mount, the public will "demand" immediate withdrawal, i.e., U.S. casualties result in an inexorable demand to withdraw precipitously from a U.S. military commitment.[1] The counter-conventional wisdom has it that U.S. casualties in fact inflame the American public, resulting in a "demand" for escalation to a "decisive victory."[2] As will be seen from an examination of the data for the Korean and Vietnam wars and Somalia, neither of these extreme views is accurate.

---

[1]Such a view is often expressed in the media.

[2]For example, Schwarz (1994, p. 12), states: "In fact, polls in both [the Korean and Vietnam] wars show an inverse relationship between "approval" of the intervention and the public's desire to escalate to achieve decisive results."

This chapter will examine aggregate trends in sentiment for an increased or decreased commitment in Korea, Vietnam, and Somalia. The next chapter will show that differences in support and diverging policy preferences among the public generally appear to derive from polarization among partisan leaders.

THE KOREAN WAR

The data on policy preferences in Korea—what Mueller refers to as "escalation" and "withdrawal"—are quite treacherous and difficult to analyze because of changing question structure and wording, inter-vening events, and other factors.[3]

To take question structure and wording as but two examples, when respondents were offered two options—a choice between pulling out of Korea or continuing the fighting—this stark choice generally resulted in the continuation option beating the pullout option: Support for pulling out ranged from 12 to 36 percent, while support for continuing the war ranged from 48 to 77 percent.[4] When faced with a three-way choice between pulling out, continuing, or escalat-ing, however, withdrawal sentiment ranged between 13 and 30 per-cent, support for continuation of the war ranged from 22 to 46 per-cent, and support for escalation ranged from 30 to 44 percent.[5]

---

[3]As Iklé (1991), pp. 39–42, notes, the concept of "escalation" is itself quite vague, often leading to conflicting interpretations.

[4]See Mueller (1973), Table 4.4, for a rather exhaustive compilation of these questions. The exception was a Gallup poll in early January 1951 that had biased wording sug-gesting China had "forces far outnumbering the United Nations troops there," and offering a choice between "pulling our troops out as fast as possible" and "keeping our troops there to fight these larger forces." In effect, this question was offering a with-drawal or a slaughter. This question found 66 percent willing to "pull our troops out of Korea as fast as possible."

[5]The sole exception was a May 1952 question from Roper in which 53 percent wanted to "stop fooling around and do whatever is necessary to knock the Communists out of Korea once and for all" when they were asked what should be done in Korea.

Escalation Sentiment

Mueller found that open-ended questions, while providing "a considerable bounty of escalatory sentiment," only rarely achieved majority support for escalatory options:

> Appeals to belligerent escalation had a quick appeal—unless it was suggested that such a policy might have some undesirable side effects .... Because so little data from open-ended questions are available and because of the changes in coding categories used by the survey organizations, little can be gleaned from these questions about trends over the course of the war.[6]

In short, he found no clear evidence of majority or growing sentiment for escalation during the war.[7] Figure 3.1 plots the percentages supporting escalation options during the Korean War.[8]

The figure lends support to Mueller's observation that, in both the Korean and Vietnam wars, "with considerable gyrations, between 20 and 50 percent are generally found to take the escalation option when several are presented."[9]

Withdrawal Sentiment

Figure 3.2 plots the percentages supporting various withdrawal options that were offered during the Korean War. In his analysis of these data, while noting the sensitivity of responses to differences in question wording, timing, and presidential policy, Mueller found evidence of *gradually increasing* sentiment for withdrawal in Korea:

---

[6]Mueller (1973), p. 9.

[7]There is also no evidence of growing support for escalation in Korea among those who disapproved. By contrast, Schwarz (1994, p. 13) argues that: "Over time, more and more respondents preferred the escalation option," and suggests growing support for escalation among those who disapproved: "Frustration, however, led not to cries to withdraw but to a desire to escalate."

[8]The figure plots the percentage of respondents supporting escalatory responses in Mueller (1973), Table 4.4. The points are responses to options that were offered only once, while the lines represent questions asked on more than one occasion See Mueller (1973), Table 4.4, for the precise wording of the questions.

[9]Mueller (1973), p. 102.

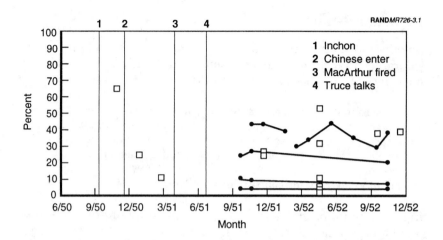

Figure 3.1—Support for Escalation in Korea

[The data] suggest that about 15 to 35 percent of the American population generally favored withdrawal from the Korean War at various times ... the sentiment for withdrawal gained somewhat in
popularity over the course of the war, although it certainly never
reached overwhelming proportions.[10]

Thus, support for escalation in Korea ranged from 20 to 50 percent,
depending on question wording and other factors, while 15 to 35
percent supported withdrawal. What are we to make of preferences
in Korea?

An Unhappy Equilibrium

The data suggest that continuation of the war, not escalation or
immediate withdrawal, came to be the de facto preference. The reason for this is that each extreme option was defeated by a coalition of
those in the center and the other extreme: Those who preferred

---

[10]Mueller (1973), p. 98.

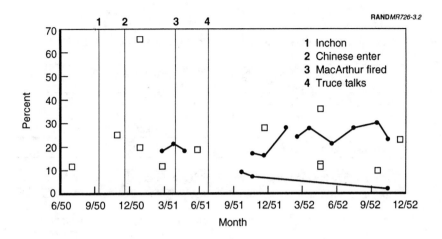

Figure 3.2—Support for Withdrawal During the Korean War

withdrawal were outnumbered by those preferring either continuation or escalation, while those who wanted to escalate were outnumbered by those who would either withdraw or continue the war.[11] In fact, Table 3.1 presents illustrative cross-tabulated data from the NES showing that, by November 1952, continuation of the war was preferred to escalation by a plurality of the total sample and by both those who supported and those who opposed the war.[12]

In short, the weight of evidence on the Korean War does not suggest a growing preference for "escalation to victory" *or* a majority preference for withdrawal. A majority consistently supported continuing

---

[11]While none of the questions posed a two-way choice between continuation and escalation, the available data suggest that escalation would probably have lost if such a choice had been offered. In only one case did a three-way question show higher support for escalation than did a combination of those who would pull out with those who would continue the war. In all other cases, less than a majority supported escalation. In the parlance of social choice theorists, it appears that continuation of the war was a Condorcet winner in pairwise voting for the three options.

[12]Mueller (1973), p. 67, presents a cross-tabulation of NORC data from November 1952 that also fails to show majority sentiment for escalation among either those who supported or opposed the war. In short, a preference for continuation over either escalation or withdrawal appears to be a reasonably robust finding.

Table 3.1

The Mistake Question and Policy Options, November 1952

Do you think we did the right thing in getting into the fighting in Korea two years ago or should we have stayed out?		
	Support, right thing	Oppose, stayed out
Percentage of total sample	37	39

Which of the following things do you think it would be best for us to do now in Korea?			
	Percentage of total sample	Support, right thing	Oppose, stayed out
Pull out entirely	9	2	18
Keep trying to get a peaceful settlement	43	49	42
Take a stronger stand and bomb Manchuria and China	35	45	34
No opinion	10	3	7

SOURCE: NES.

the war until a negotiated settlement and an orderly U.S. withdrawal—especially to include the recovery of U.S. POWs—could be achieved; only a minority supported escalation; and only a small (but slowly growing) percentage supported the withdrawal option.[13] Some Republican leaders had earlier gone on record opposing a settlement that left Korea divided.[14] However, other American political leaders and the public supported the armistice: In July 1953, NORC found that the public preferred the armistice to continued fighting by a five-to-one margin.[15]

---

[13]These differences will be described in more detail in the next chapter.

[14]Although he ultimately backed the Korean truce, Senator Taft stated his opposition in April 1953 to a settlement that would result in the partition of Korea. *Facts on File Five Year Index, 1951–1955*, p. 335.

[15]Mueller (1971) reports that NORC Survey No. 348 found that 75 percent preferred the armistice, while 15 percent preferred to continue fighting.

Conclusions

As long as President Truman met three conditions, he was able to continue the war until a negotiated settlement could be achieved:

- pursuit of a negotiated settlement that would allow an orderly withdrawal following the release of U.S. POWs

- reliance on a strategy to minimize the costs of the U.S. commitment until that negotiated settlement could be achieved[16]

- continued willingness to accept the considerable political costs of continuing the war, including the willingness to sacrifice his presidency, if necessary.

The result of pursuing the middle course was an unhappy equilibrium in which a majority grudgingly preferred continuation of the war over either withdrawal or escalation and in which most were unwilling to quit the war until U.S. POWs had been returned.

THE VIETNAM WAR

Like Korea, the Vietnam War was a long, drawn-out conflict that occasioned frustration and incompatible policies for war termination.

Figure 3.3, showing the evolution of policy preferences toward the war, is representative of time-series public opinion questions that offered three policy options for Vietnam: increasing, maintaining, or reducing the level of commitment to the war. The figure presents the results of a three-way question that NES asked every two years from 1964 to 1970.

The figure suggests that escalation sentiment peaked perhaps in 1966 and 1967 and declined thereafter, while withdrawal sentiment

---

[16]According to Hosmer (1985), p. 66, General Matthew B. Ridgway "went over to active defense in November 1951, eschewing any further major ground offensives in the war," thereby adapting war strategy to minimize casualties. Goldhamer (1994), pp. 138–139, makes a similar point, suggesting that, by late October 1951, the Joint Chiefs of Staff and Far Eastern Headquarters wanted to avoid military activity, thereby avoiding "wasted lives."

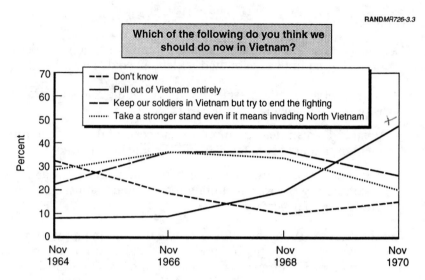

RAND*MR726-3.3*

Figure 3.3—Policy Preferences on Vietnam, 1964–1972

increased from 1966 to 1970, although never achieving majority
levels of support. This finding seems generally to be supported by
the examination of questions on escalation and withdrawal that will
next be presented.[17]

---

[17]Judged by a similar result in other three-way questions that offered escalation,
continuation, and withdrawal options, this seems to be a robust finding. Gallup and
NORC asked a question five times from August 1966 to April 1968 (three of which are
included as the R1–R3 series in Mueller, 1973, Table 4.5):

> Just from what you have heard or read which [one of the statements on the
> card, of these statements] comes closest to the way you, yourself, feel about the
> war in Vietnam? A. The U.S. should begin to withdraw its troops; B. The U.S.
> should carry on its present level of fighting; C. The U.S. should increase the
> strength of its attacks against North Vietnam; D. No opinion.

The percentage preferring withdrawal increased rather steadily from 17 percent in
August 1966 to 39 percent in April 1968, while the percentage who wanted to increase
the strength of attacks decreased from 55 percent to 37 percent. Gallup, August 18–23,
1966, October 1966, August 24–29, 1967, October 1967, and NORC/SRS, April 1968. A
series from the *Washington Post* (U1–U3 in Mueller, 1973, Table 4.5) suggests a rather

It also shows that continuing the war until a negotiated settlement could be achieved generally remained the most popular option, although by 1970 the percentage who wanted to pull out and the percentage who wanted to continue were roughly equal.[18]

Escalation Sentiment

Mueller's (1973) analysis of public opinion data on escalation sentiment in Vietnam led to the conclusion that the Vietnam War generated escalatory sentiment to about the same degree as in Korea, ranging between about 20 and 50 percent, depending on question wording and structure, presidential policy at the time, and other factors. Figure 3.4 summarizes the data he presented on support for escalatory options.

Judging from the few time series (questions with identical wording asked over time, represented by the lines), the figure seems to confirm that escalatory sentiment may have peaked in 1966 or 1967 and declined thereafter, especially after the Tet offensive of early 1968.[19] Support for the bombing of North Vietnam may also have peaked in 1966.[20]

---

similar finding for July–September 1967, with withdrawal sentiment increasing from 24 to 37 percent and escalation sentiment remaining steady at about 20 percent.

[18]In fact, among those who disapproved of the war, there was growing sentiment for withdrawal and declining sentiment for escalation. This suggests the thin evidentiary basis of Schwarz' (1994) contention that::

> As was the case with the Korean War, as the conflict continued, as casualties mounted, and as 'disapproval' of the commitment grew, an increasing number of those polled found escalation the most attractive option." Schwarz (1994), p. 14.

[19]The impact of Tet on escalation sentiment is also suggested by the fact that those who identified themselves as "hawks" (i.e., wanted to increase the level of commitment) rallied by about eight points briefly at the time of Tet, but then fell nearly 20 points. See Gallup (1972), pp. 2105–2106, 2124–2125.

[20]A polling result that supports this view is Gallup's finding that 49 percent of those polled in July 1967 opposed a proposal that would send 100,000 more troops to Vietnam, while 40 percent supported the proposal. Gallup, July 13–18, 1967. By March 1968, 51 percent said they would approve if the government decided to stop the bombing and fighting and gradually withdraw from Vietnam; in April 1968, 64 percent approved of President Johnson's bombing halt. Gallup, March 15–20, 1968, and April 4–9, 1968.

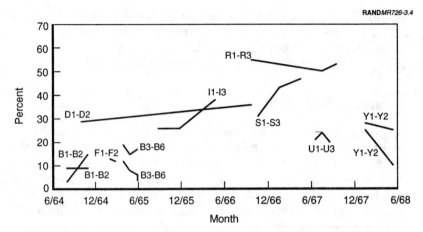

RAND*MR726-3.4*

NOTE: The labels represent polling questions found in Mueller (1973), Table
4.5, which use a wide variety of question wordings. There are two lines each for
B1–B2, B3–B6, and Y1–Y2 because each question offered two separate military
escalation options. Readers interested in the differences will be rewarded by
examining the table in Mueller's book.

Figure 3.4—Support for Escalation in Vietnam

Withdrawal Sentiment

While noting that questions suggesting a policy of gradual withdrawal tended to receive much higher levels of support than questions that implied an immediate withdrawal,[21] Mueller suggested
that attitudes on withdrawal during the Vietnam War were quite
similar to those in the Korean War:

> [I]t appears that the sentiment for withdrawal for the two wars was
> roughly the same. Particularly for the pre-1969 period, support for
> withdrawal from Vietnam in the blunt Korean-style question ("pull
> our troops out") ranges upward, as in Korea, to a maximum of

---

[21]For example, in October 1967, 71 percent of those polled approved of an option
Gallup presented in which U.S. troops would be brought home as Vietnamese troops
were trained. Gallup, October 1967. Fifty-six percent told Gallup in March 1968 that
they would approve of a government-led de-escalation and withdrawal. Gallup,
March 1968.

around 35 percent ... it does seem that the wars in Korea and Vietnam stimulated popular yearning for the policy option of with-drawal to much the same degree.[22]

This finding is generally confirmed by the data in Figure 3.5, which presents various series showing increasing sentiment for withdrawal. In Vietnam, as in Korea, sentiment for immediate withdrawal seems to have grown but never to overwhelming levels. If neither extreme option—escalation or immediate withdrawal—came to be the domi-nant preference during the Vietnam War, what was the outcome?

A Return to the Unhappy Equilibrium

The answer seems to be that, like Korea, pluralities or majorities consistently preferred to achieve a negotiated settlement and orderly withdrawal from Vietnam only after U.S. POWs were recovered, and the extreme options of escalation or immediate withdrawal could be consistently defeated by a coalition of those in the center and those at the other extreme.[23]

Figure 3.6 presents data from a question NES asked in November 1968, asking the respondent to identify his preferred policy on a con-tinuum.[24] By November 1968, there was indeed polarization to the

---

[22]Mueller (1973), pp. 101–102. President Nixon seems to have pursued a strategy of using announcements of reductions in U.S. forces to buy time. In a private communi-cation, George (1995) cites Holl (1989). In fact, Page and Shapiro (1992, pp. 237–239) found a statistical relationship between the percentage of respondents to the Harris poll saying that withdrawals were "too slow" and the actual withdrawal rate one month later, suggesting that Nixon's instinct was basically correct.

[23]These extremes were also represented in the Congress. Dean Rusk reports that, before March 1968, the Johnson administration's main opposition came from those who wanted to escalate the war, not those who wanted to withdraw. See Garfinkle (1995), p. 511. Nevertheless, a growing number of Democratic senators became more vocal about their preference for de-escalation and even undertook a number of efforts to cut off funding. See Zelman (1984).

[24]The question the NES asked was:

There is much talk about "hawks" and "doves" in connection with Vietnam, and considerable disagreement as to what action the United States should take in Vietnam. Some people think we should do everything necessary to win a com-plete military victory, no matter what results. Some people think we should withdraw completely from Vietnam right now, no matter what results. And, of

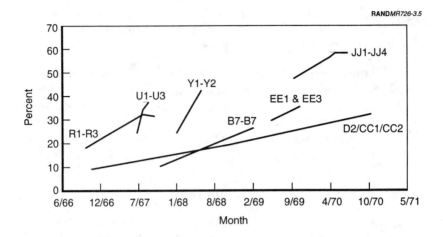

Figure 3.5—Support for Withdrawal from Vietnam

extremes of complete military victory and immediate withdrawal, but these extremes did not command the support of a very large segment of the population—no more than one in five at each end of the continuum. In November 1968, then, the data show that the most common preference was a centrist one—continuation of the war, neither escalating nor withdrawing, until peace and an orderly withdrawal could be negotiated.[25]

In short, Schwarz's (1994) thesis about a protean public increasingly demanding escalation to victory in Vietnam is supported neither by

---

course, other people have opinions somewhere between these two extreme positions. Suppose the people who support an immediate withdrawal are at one end of this scale (show card to respondent) at point number one. And suppose the people who support a complete military victory are at the other end of the scale at point number seven. At what point on the scale would you place yourself?

[25]Although a plurality of 30 percent could be found in the centrist position, slightly more respondents—31 percent—were in one of the two extreme positions, and those at the extremes were probably more vocal in their preferences. I am indebted to Miroslav Nincic for this point.

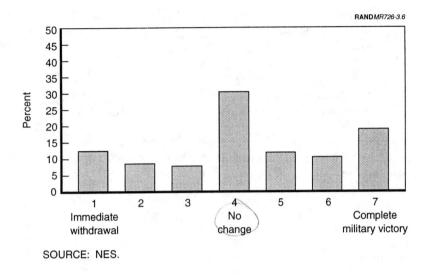

RANDMR726-3.6

SOURCE: NES.

Figure 3.6—Policy Preferences on Vietnam, November 1968

Mueller's analysis nor by the weight of the available evidence.[26] But neither is the belief justified that declining support led to majority support for immediate withdrawal. The truth is that only minorities occupied each extreme, while pluralities or majorities ("the Silent Majority") occupied a centrist position.

While support for *immediate* withdrawal from Vietnam was somewhat muted, majorities could frequently be found for a *gradual* or *orderly* withdrawal, even if this involved the consequent loss of South Vietnam. In April 1971, when Harris asked:

> [i]f the reductions of U.S. troops continued at the present rate in Vietnam and the South Vietnamese government collapsed, would you favor or oppose continuing the withdrawal of our troops?

---

[26]Schwarz argues that "Vietnam presents an even sharper picture of the public passion for escalation as conflicts continue." Schwarz (1994), p. 13.

60 percent favored continuing the withdrawal.[27] When the safety of American POWs was mentioned, however, even larger percentages refused to withdraw. In May 1971, when ORC asked:

> Would you favor withdrawal of all United States troops by the end of 1971 even if it threatened the lives or safety to United States POWs held by North Vietnam?

fully 75 percent *opposed* withdrawal.[28] As the war drew to a conclusion, the settlement that most Americans preferred appears to have been entirely bound up with the issue of getting U.S. POWs back.[29]

Conclusions

The re-examination of Mueller's data and findings and the examination of a great many other public opinion questions lead to the clear conclusion that in Vietnam, as in Korea before it, the leaders and the public became polarized over war aims and policy, leaving an unhappy equilibrium in which the president was grudgingly given support to negotiate a settlement that would assure an orderly withdrawal including the release of U.S. servicemen held as POWs.[30] Also as in Korea, the price of this grudging support for the war was accepting the mounting political costs of its prolongation, while minimizing the rate at which the United States continued to incur casualties until a negotiated settlement that would return U.S. POWs could be achieved.[31]

---

[27]Twenty-six percent opposed, and 14 percent were not sure. Harris, April 1971.

[28]By comparison, 11 percent favored withdrawal, and 14 percent had no opinion. ORC, May 1971.

[29]I am indebted to Professor John Mueller for this point.

[30]On this point, see also Mueller (1973), pp. 97–98, and Mueller (1989), pp. 180–181.

[31]By late 1969, when also offered a choice of immediate withdrawal and orderly withdrawal, a plurality of 30 percent supported "staying on in Vietnam until a satisfactory settlement can be reached, but fight the war in such a way as to keep American casualties as low as possible." By comparison, 25 percent supported withdrawal of U.S. troops as quickly as they could be moved out, and 24 percent supported withdrawing U.S. troops gradually, "within the next year or so." In short, by late 1969, a U.S. withdrawal—immediate or gradual—was generally preferred to continuation. Gallup, December 4–9, 1969. See Lorell et al. (1985) on the role of casualties and public opinion on presidential policy during the Vietnam War, and Hosmer (1985), pp. 63–

SOMALIA

Attitudes about Somalia are in some respects similar to, but in other respects different from, those about Korea and Vietnam.

Even before the deaths of the Rangers in Mogadishu, public support had fallen to only about four out of ten, and the Congress was both increasingly vocal in its opposition to the mission and already moving to limit the operation. Following the Ranger deaths in Mogadishu, majorities supported the compromise between the White House and Congress that called for an orderly withdrawal from Somalia at the end of March 1994, while only minorities favored either an immediate withdrawal or an escalated commitment.

Nevertheless, majorities *were* willing to support attacks on Aidid and his forces if these attacks were judged necessary to secure the safe release of U.S. servicemen held hostage; they approved of additional forces but only if it was indicated that these forces would be used to better defend forces already in place; and they were also willing to support attacks to punish Aidid for the deaths of the Rangers, so long as these attacks did not impede an orderly withdrawal. In short, the balance of the evidence does not support the proposition that majorities wanted an immediate withdrawal, but neither does it support the proposition that the public would have supported "escalation" in the sense of increasing the level of commitment to Somalia. As described earlier, only a minority felt that Somalia was important enough to warrant such an increased commitment.

The deaths of the Pakistani peacekeepers in June 1993 and the deteriorating situation thereafter led political leaders, the media, and the public to pay higher levels of attention to Somalia.[32] In late June 1993, 50 percent thought the United States had a responsibility to do something about the situation, and United Nations efforts to capture Aidid received support from two out of three of those polled.[33] The public was nevertheless divided over the likely consequences of such

---

70, on the role of casualty-minimizing strategies in maintaining support for a negotiated settlement.

[32]Gallup's June 18–21, 1993 poll found 63 percent following events in Somalia very or somewhat closely; in September, Times Mirror found that 62 percent were still following Somalia closely.

[33]CBS News/*New York Times*, June 21–24, 1993.

a response: A plurality of 42 percent thought the United States would get bogged down if it undertook attacks against Aidid, while 41 percent approved.[34]

By mid-September, the public had become more concerned about the direction being taken in Somalia: While 69 percent thought the development of a peaceful solution should be a very or somewhat important U.S. goal, 57 percent thought the United States should stop active military involvement.[35] Fifty-two percent thought that the United States was too deeply involved in Somalia.[36] By contrast, only 43 percent approved of the presence of U.S. troops in Somalia.[37] Delivering food was preferred to disarming warlords by a three-to-one margin.[38]

The deaths of nearly 20 U.S. servicemen in the firefight in Mogadishu on October 3–4, 1993, came as a shock to the American body politic, resulting in dramatic increases in congressional and other political activity and media reporting. Where the number of American deaths due to hostile action had nearly tripled between May and September 1993, that number nearly tripled again in a single incident.

As in the cases of Korea and Vietnam, neither those who have argued that the public wanted immediate withdrawal nor those who argued that the public in fact preferred an increased commitment in Somalia are quite correct.[39] A majority of the public was typically averse to either an immediate withdrawal or an increase in the level of commitment to achieve broader policy objectives.

Just as in Korea and Vietnam, there was polarization between those who would do more (about one in four) and those who would reduce the commitment by withdrawing (about two out of three). Unlike Korea and Vietnam, however, where U.S. interests had resulted in strong support for the centrist position of continuing until a negoti-

---

[34]CBS News/ *New York Times*, June 21–24, 1993.

[35]Gallup/CNN/ *USA Today*, September 10–12, 1993.

[36]NBC News/ *Wall Street Journal*, September 10–13, 1993.

[37] *Time*/CNN, September 23–24, 1993.

[38] *Time*/CNN, September 23, 1993.

[39]Kull and Ramsay (1993), p. 4, argue that the public opinion data show "majority sentiments in support of increased involvement, at least in the short run."

ated settlement could be reached, U.S. stakes in the Somalia intervention did not justify maintaining or increasing the commitment to most members of the public; the clear preference was for an orderly (but not an immediate) withdrawal.[40] A reasonably representative polling result on questions that offered withdrawal and escalation options is found in Figure 3.7. It shows that a plurality wanted to withdraw immediately, and one in four preferred an orderly withdrawal (for a total of about two out of three preferring withdrawal); only about one in four wanted to send more troops to Somalia. In fact, majorities rather consistently expressed a willingness to withdraw even if the situation in Somalia deteriorated after the withdrawal.[41] The story gets a bit more complicated from here, however.

Although withdrawal from Somalia was consistently the preferred option for most of those polled, majorities refused to withdraw until U.S. servicemen held hostage had been safely returned.[42] The data also show that majorities of the public were willing to support additional military measures, but only in a limited context. Additional forces were supported, for example, when it was suggested that they would be used to better protect troops already in Somalia, or would be used to secure the release of U.S. service personnel held hostage by Aidid.[43]

Complementing a majority's unwillingness to withdraw until U.S. servicemen held hostage were returned, the use of force was sup-

---

[40]Of the 30 or so polls on Somalia that offered a withdrawal option after the October firefight in Mogadishu, majorities rather consistently supported withdrawal, and pluralities often supported immediate withdrawal. In only one poll—that done by ABC News on October 7, 1993—did there seem to be *majority* sentiment for an immediate withdrawal, however. Data are available from the author.

[41]ABC News, October 7, 1993; Gallup, October 8–10, 1993; ABC News, October 5, 1993 and October 7, 1993.

[42]See CBS News, October 6, 1993 and October 6–7, 1993; NBC News, October 6, 1993; *Time*/CNN, October 7, 1993; and Gallup, October 8, 1993.

[43]ABC News, October 5, 1993; Gallup, October 8–10, 1993; and NBC News, October 6, 1993 and October 8, 1993. When simply asked whether they approved of sending additional forces without specifying a purpose that was instrumental to an orderly withdrawal, however, majorities consistently opposed sending additional forces. Gallup/CNN/*USA Today*, October 5, 1993; ABC News, October 7, 1993; *Washington Post*, October 7–10, 1993.

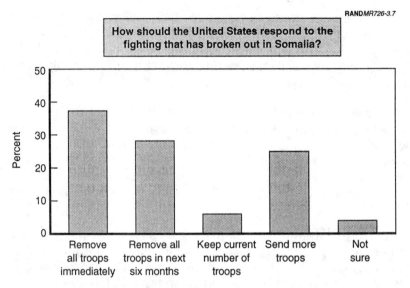

RAND*MR726-3.7*

SOURCE: *Time* CNN/Yankelovich (10/7/93).

Figure 3.7—Policy Preferences on Somalia, October 7, 1993

ported if negotiations failed to release the U.S. servicemen.[44] The principle that any attack on U.S. troops required a stronger U.S. military response from the United States also won support.[45] Support for capturing Aidid ranged from 51 to 71 percent, depending on timing and question wording.[46] But when questions implied that efforts to capture and punish Aidid might delay a U.S. withdrawal, support for the option evaporated.[47] In short, a rather consistent

---

[44]ABC News, October 5, 1993 and October 7, 1993.

[45]*Time*/CNN, October 7, 1993.

[46]Gallup/CNN/*USA Today*, October 5, 1993; ABC News, October 5, 1993, and *Time*/CNN, October 7, 1993.

[47]CBS News, October 6, 1993 and October 6–7, 1993; Gallup, October 8–10, 1993. A conjecture is that majorities were willing to capture *or* punish Aidid so long as it did not delay an orderly withdrawal—two out of three were concerned that sending more troops would just get the United States more deeply involved in Somalia. CBS News, October 7, 1993 and October 18–19, 1993. This interpretation is consistent with cross-tabulated results of Gallup's October 5, 1993 poll, that show that 63 percent of those

majority found the benefits of an increasing commitment to Somalia not worth the additional costs.[48] Majorities supported actions that were instrumental to an orderly withdrawal but were more than happy to punish Aidid if it would not hinder withdrawal or result in additional U.S. casualties.

Revisiting the Myth of the "CNN Effect"

The reader will recall that the data fail to support the argument that heightened media reporting on Somalia preceded, much less drove, presidential decisionmaking. There is also no evidence supporting another bit of conventional wisdom on Somalia: that media images of the mistreatment of dead U.S. servicemen were responsible for the public's desire to withdraw from Somalia (see Table 3.2).[49]

A tabular and statistical analysis of public opinion data reveals that a majority of both those who saw the images and those who did not favored withdrawal from Somalia. While the images did not affect the *direction* of the public's preferences, they do however appear to have affected the *degree* of these preferences: Those who saw the images were somewhat more inclined to support immediate withdrawal than those who did not, although the largest percentage among both groups favored immediate withdrawal.

In short, there is again cause to be skeptical of arguments about the impact of the media on public opinion on Somalia: The media appear to have followed objective events, conditions, and presidential decisionmaking, and "effects" were at best modest, apparently reinforcing preexisting preferences.

---

who wanted to withdraw immediately supported capturing and punishing Aidid, as did 71 percent of those who wanted to withdraw gradually. Gallup's October 8–10, 1993 poll shows that only 18 percent of those who wanted to withdraw immediately and 24 percent of those who wanted to withdraw over six months supported keeping troops in Somalia until Aidid was captured and punished.

[48]Thus, the hypothesis that casualties led to a desire for an *increased* commitment to Somalia is not substantiated.

[49]The reader will recall that only about four in ten in September 1993 approved of the Somalia operation; support fell a further ten points after the firefight in Mogadishu.

Table 3.2

Viewership of Televised Images and Policy Preferences

Next, I'd like you to think back to the news you may have seen over the last days. By any chance have you, yourself, happened to see the actual news photo of either a U.S. soldier's corpse being dragged through the streets of Somalia, or the television footage of a captured U.S. pilot being interviewed?

	Yes	No
Percentage of total sample	58	39

In your view, what should the United States do now in Somalia:

	Yes (percent)	No (percent)
One: Withdraw U.S. troops right away	50	33
Two: Gradually withdraw U.S. troops	23	30
Three: Keep U.S. involvement the same	6	10
Four: Increase U.S. military commitment	18	20

SOURCE: Gallup/CNN/ *USA Today* (October 5, 1993).

Conclusions

The evidence on Somalia does not suggest that the public and the government responded largely to televised imagery, that the majority of the public had desired that the United States "escalate to victory" as a result of casualties, or that a majority "demanded" an immediate withdrawal. In fact, Somalia represents another case in which the historical record suggests a more sensible and subtle response to increasing casualties and declining support: A plurality or majority has typically rejected both extreme options of escalation and immediate withdrawal and has remained unwilling to withdraw until a negotiated settlement and orderly withdrawal—including the return of U.S. servicemen—could be concluded.

CHAPTER CONCLUSIONS

Opposition to a war or military operation can come either from members of the public who prefer a decreased commitment (deescalation or withdrawal) or from those who believe that more should be done to achieve a successful outcome, i.e., those who desire an increased commitment (or escalation). In its most extreme form, some have argued that casualties and declining support have typically led to increasing demands for immediate withdrawal, while others have argued that casualties and declining support have led to inexorable demands for escalation to victory.

The data appear to contradict both extreme positions, while being broadly consistent both with other past RAND work and with work by other scholars that demonstrates the importance of leadership, and objective events and conditions, in public support.   More specifically, because of the importance of the interests, pluralities or majorities of the public during Korea and Vietnam grudgingly supported continuation of each war until a settlement and an orderly withdrawal could be achieved and supported temporary escalation to break diplomatic deadlocks, if the costs were reasonable.   In Somalia, majorities of the public in October 1993 were also unwilling to be forced out of Somalia in a precipitous fashion (i.e., without recovering U.S. servicemen held hostage).   However, they were also unwilling to stay longer than the six months the president negotiated with the Congress for accomplishment of an orderly withdrawal; the stakes simply did not warrant such a commitment.

When the perceived benefits are low or success is particularly elusive, the settlement that Americans prefer is often entirely bound up with the rather limited issue of getting prisoners or hostages back; once accomplished, there may be little to warrant a continued presence. The case of Somalia shows, however, that, even when majorities prefer withdrawal, they may often be willing to support punitive strikes, as long as an orderly withdrawal is not delayed.

Individuals clearly may differ greatly in their evaluations of the benefits of an operation, expectations of success and failure, and willingness to make trade-offs between benefits and costs. We might also expect them to differ in their optimism about the prospects that escalation will lead to success at low cost, which would figure in support for escalation. But why might individuals differ on these questions?

As will be seen in the next chapter, there is good reason to believe that differences in beliefs and preferences among the public have followed from and paralleled differences among political and other opinion leaders.

Chapter Four

LEADERSHIP CONSENSUS AND DISSENSUS

The case studies in the previous chapters examined a number of factors associated with support for U.S. military interventions, focusing on assessments of the stakes, objective events, and conditions. The role of leadership in public support and the impact of consensus or dissensus among leaders, however, were given only passing treatment. This chapter will therefore address this important issue through an examination of four of the cases: the Korean War, Vietnam War, Gulf War, and the U.S. intervention in Somalia.

To put this chapter in perspective, the argument that is offered is a somewhat stylized one that focuses on the role of leadership in the "democratic conversation," at the expense of the public influences on the leaders. This perspective represents a conscious effort to bring political leaders back into the picture, where they belong, but is in no way meant to imply that the democratic conversation is simplistic, top-down, and unidirectional, solely involving communications from leaders to a pliant public. Rather, the perspective relies upon the insights that followers and leaders have partisan ties of varying strength, that they share many of the same ideological or political leanings, and that members of the public accordingly can find reliable cues on complex issues by looking for the positions of the leaders they find most credible. In short, individuals ultimately choose which arguments are most credible but use a shortcut that reduces their information-gathering costs.[1]

---

[1]See Downs (1957), pp. 245–247.

LEADERSHIP AND FOLLOWERSHIP ✓

The way public opinion is sometimes treated in political and media circles suggests that many in those circles believe that it measures the strongly held positions of a rather autonomous, strong-willed ✓ and decisive public. However, a great deal of research suggests that public opinion on policy issues responds not only to objective events and conditions (as was discussed in the last two chapters) but also responds in a predictable way to political leadership.[2] ✓

A number of works have examined the impact of presidential leadership on public attitudes or have described the conditions under which presidents may mobilize support for policies.[3] Other work has examined public attitudes toward issues on which partisan leaders have disagreed and found that these divisions polarize the public in a ✓ predictable way.[4] A growing body of this research has examined the relationship between leadership and public opinion on military operations.[5]

This research suggests that the diffusion of mass attitudes is perhaps best explained through a combination of aggregate- and individual-level processes. At the aggregate level, political debates that are given media attention result in a mix of pro and con positions being reported to members of the public.[6] At this point, two individual processes take over. One describes differences in the likelihood of becoming aware of these pro or con positions, and the other

---

[2]See, for example, Page and Shapiro (1992) and Kagay (1992).

[3]See Mueller (1973), pp. 196–241); Page, Shapiro, and Dempsey (1987); Brody (1991), pp. 45–78; and Burbach (1994).

[4]See Gamson and Modigliani (1966); Chong, McClosky, and Zaller (1984); and Zaller (1992).

[5]Cantril (1947), for example, analyzed public opinion data from the Second World War and showed that in many cases differences in knowledge levels produced differences comparable to partisan differences. Belknap and Campbell (1951–1952) extended this insight by exploring the impact of a combination of partisanship and knowledge on attitudes toward the Korean War. More recently, Zaller (1992, 1993) has examined the role of political knowledge and partisanship or ideology in the Vietnam and Gulf wars.

[6]Aggregate-level factors that may affect the probability of reception include the amount of attention political leaders and experts devote to an issue, the intensity of the messages from leaders, and the extent or level of media reporting devoted to these messages.

describes which positions are most likely to be accepted, and which rejected.

Many readers may be familiar with the notion that members of the public are either "attentive" or "inattentive." Over the last 30 years, this dichotomous framework has been elaborated by treating the public's political awareness on a continuum rather than in terms of this simple dichotomy.[7] Although the operationalization can vary— indexes of media usage, political knowledge, or sophistication, for example—the basic premise of this research is that individual differences in political interest affect the probability of becoming aware of new information when it is reported in the media.[8] In short, some will quickly become aware of new developments (or pro or con messages from political leaders), while others will become aware only much later.

Once having become aware of the messages in the leadership debate, personal partisan (or ideological) leanings determine which messages individuals decide to accept or reject.[9] Thus, the most aware and partisan (or ideological) members of the public are first to receive and accept (or reject) partisan-coded messages from their opinion leaders, while the least aware and partisan are last to receive—and least likely to accept or reject in a consistently partisan

---

[7]See Key (1961), pp. 282–285 for a discussion of the attentive and inattentive publics.

[8]That is, aggregate-level factors that may affect the probability of reception include the intensity of the messages from leaders and the extent or level of media reporting devoted to these messages. There seems generally to be agreement that knowledge of and the ability to reason about politics are the measures of real interest. There are a number of different operationalizations of this concept, e.g., measures of political awareness, knowledge, or sophistication. Self-ratings on media usage or information level and on education are also positively correlated with these concepts, although these measures are crude approximations at best. For more in-depth discussions of the strengths and limitations of these various measures, see Neuman (1986), pp. 51– 81, who favors a measure of "political sophistication," and Zaller (1992), who favors a measure of "political awareness."

[9]Zaller (1992) calls this the Receive-Accept-Sample (RAS) model—individuals receive new information, decide whether to accept it, and then sample at the moment of answering public opinion questions. Ideology and "hawkishness" or "dovishness" have also been substituted.

fashion—messages on the basis of party loyalty or ideological affinity.[10]

For present purposes, the result is that, when bipartisan leadership consensus exists on a military operation, because there are few negative messages available to the public, there is typically also consensus in the public.[11] When the leadership is characterized by dissensus, however, the public also tends to become polarized. In short, leadership consensus or dissensus can figure prominently both in building and maintaining support for U.S. military operations and in influencing preferences on policy and strategy.

As mentioned earlier, the foregoing description of opinion leaders is not an unflattering suggestion that the people are automatons mindlessly following their leaders and are therefore easy marks for manipulation. In fact, public opinion often displays a great deal of inertia or "viscosity," remaining somewhat resistant to the persuasive efforts of leaders.[12] The reasons for this may be complex and varied—in some cases because the issue is not very salient, in other cases because the persuasive arguments lack credibility, in still others because presidents have been challenged or criticized by other leaders.

In any case, the key point is that the individual-level factors ultimately determine which arguments or positions an individual will accept and which he will reject. In short, where the public may be constrained to consider the policy options the leaders offer, the leaders are also constrained in their efforts to mobilize support, because individuals ultimately decide for themselves which arguments are most credible.[13]

---

[10]Other factors, such as ideology or hawkishness, may sometimes play the same role as partisanship in discriminating between messages, depending on the availability of leaders who advocate these viewpoints.

[11]In this light, the so-called "rally 'round the flag" effect may be partially explained by the initial absence of leadership critics—and the presence of favorable commentary—when a president uses force.

[12]Viscosity is V.O. Key's term. See Key (1961), Chapter 10.

[13]Research on the so-called "rally 'round the flag" effect, for example, suggests that rallies in presidential approval are typically rather small—an increase of about 3–4 percentage points—and rather short lived. See Brody (1991) and Burbach (1994). The

One final point before returning to the case studies. Some believe that the media are more important in forming public opinion than political leaders. While the media may exercise some influence by "framing" issues or choosing the viewpoints presented, a compelling case has yet to be made that the media are doing anything when they report on military operations other than documenting the ebb and flow of military and political activity, as interpreted by the viewpoints that can be sampled in mainstream political or expert circles.

In fact, a great deal of evidence suggests that media reporting levels on an issue are often related to objective events and conditions, and that the media rely extensively on the opinions of political leaders and experts when interpreting these events and conditions for their audiences.[14] It is, furthermore, not at all clear that past failures to relate media reporting levels on military operations to objective events and conditions do not have more to do with conceptual and measurement problems than with the absence of a relationship between the two.[15] Put another way, since it is much easier to measure media reporting levels than political and military events and conditions, we need to be very careful in imputing effects to the thing that can easily be measured and ignoring the thing that is harder to measure.[16]

KOREA

Korea is the first case that provides evidence of the importance of leadership dissensus in dividing members of the public.

---

impact of televised presidential addresses on public opinion, while positive, is similarly modest.

[14]See, for example, MacKuen and Coombs (1981, pp. 81–123) and Hallin (1986).

[15]For example, while he was able to relate public concern about several other policy issues to objective events and conditions, MacKuen (1981) found that concern about Vietnam was unrelated to U.S. troop strength in Vietnam. Based upon analysis of media reporting on Somalia, there is strong reason to believe that reporting levels on military operations are more closely associated with dynamic measures of military and political activity than with such measures as troop levels.

[16]As will be seen in the discussion of Somalia, when presidential decisionmaking, media coverage, and public opinion on Somalia are carefully examined, the conventional wisdom about the "CNN effect" being at work in Somalia is not substantiated. This suggests that such arguments, often based on analysis that is casual at best, should in general be viewed with great skepticism.

Support for the Korean War

The Korean War began with strong bipartisan support from congressional leaders,[17] but fell prey to partisan criticism after the Chinese
entry into the war.[18]

Belknap and Campbell (1951–1952) confirm this observation, finding
that, by the summer of 1951, increasing partisanship had led to divisions among leaders on many issues in U.S. foreign affairs, including
Korea.[19] They furthermore found that this polarized leadership
environment was mirrored in the public, suggesting that partisan differences among leaders went a long way toward explaining follower-
ship in the war:[20]

> [T]he following facts seem clear: 1) At the time of the survey the
> public was divided on issues of foreign affairs, 2) This division was
> consistently related to party identification, 3) The positions held by
> the adherents of the two major parties reflected the positions held
> by the leadership of the two parties, and 4) Independent voters
> tended to hold positions between the adherents of the two parties.
> (Belknap and Campbell, 1951–1952, p. 608.)

They present data from the University of Michigan Survey Research
Center's minor election study of June 1951 showing that differences
in support for Korea at that time were in fact tied to partisanship and
information level (see Figure 4.1).

---

[17]Zaller (1992), p. 174. Nevertheless, the Republican party was also somewhat
divided, with more isolationist Republican leaders, such as Senator Taft, criticizing the
war as early as September 1950.

[18]According to Mueller (1973, pp. 116–117):

Partisan differences were relatively small at the beginning of the [war], presum-
ably under the influence of a sort of nonpartisan consensus at a time of national
emergency. Differences broadened considerably once the [war was] underway,
becoming entirely unambiguous after the Chinese intervention in Korea.

[19]Belknap and Campbell (1951–1952, p. 603) state that "Congress was split along party
lines on the subject of foreign policy as it had not been since the 1930's."

[20]An examination of these data also showed that, as predicted, polarization was most
striking among high-information Democrats and Republicans. The attitudes of high-
information Democrats and independents, on the other hand, were rather similar.

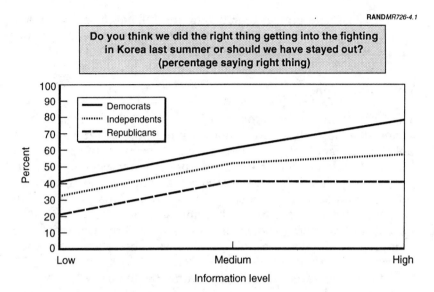

SOURCE: Belknap and Campbell (1951–1952), SRC (June 1951).

**Figure 4.1—Support for Korea by Partisanship and
Information Level, June 1951**

The figure shows that Democrats at every information level were
more likely to believe that the United States was right in going into
Korea, independents next most likely, and Republicans least likely.
And as the information level increased, the gap between Democrats
and Republicans widened: Low-information Democrats and
Republicans were separated by 19 points, while those at the highest-
information level were separated by 38 points.[21]

According to the NES of November 1952, Democrats were still more
likely than independents or Republicans to believe that getting into
the war had been the right thing to do, and the gap between
Democrats and Republicans widened after the lowest information

---

[21]A similar result obtained for approval of the actions of U.S. foreign policy leaders
and administration policy in Asia. See Belknap and Campbell (1951–1952) or Inter-
University Consortium for Political Research (1968).

level.[22] The data appear to substantiate Belknap and Campbell's finding that public support for Korea seems to have fallen prey to the same partisan divisions that existed among U.S. political leaders.

Policy Preferences

The frustrations of the setback in November 1950 and the subsequent stalemate also led to polarization among leaders over war policies. In fact, only two alternative policies were offered during Korea. President Truman and his mostly Democratic and independent supporters sought a negotiated settlement following the recapture of South Korea and were willing to continue the war until that goal could be achieved. MacArthur and his mostly Republican supporters, on the other hand, sought a widening of the war to include bombing of "privileged sanctuaries" in Manchuria and continued efforts to reunify the peninsula.[23] Partisan divisions among the leaders seem to have sowed division within the public.

The June 1951 Minor Election Study revealed that, while 72 percent supported the objective of "driving the enemy out of South Korea" (Truman's limited objective), only 42 percent supported the option of "bombing bases in China" (the option supported by MacArthur and his supporters but opposed by Truman).[24] The summer 1951 survey also found that support for Truman in his disagreement with MacArthur was associated with partisanship and information level (see Figure 4.2). The figure shows that the gap between Democrats and Republicans widened as information level increased from 30 points at the lowest level to 68 points at the highest level. Again, those most likely to be aware of partisan divisions were also most likely to be divided.

---

[22]Mueller (1973) found a similar result.

[23]In fact, following the Chinese entry into the war, MacArthur argued for one of two courses: evacuation from Korea or all-out war on China, including bombing the mainland, a naval blockade, and the use of Nationalist forces on Formosa. Congressional Quarterly (1965), p. 269.

[24]The question SRC asked was: "Here is a list of things that might be done in handling the Korean situation. Which of these things do you feel the U.S. should do and which ones shouldn't we do?" In addition to the options cited above, 52 percent supported "driving the enemy out of all Korea"; 20 percent supported "invade China"; and 19 percent supported "get out of Korea and stay out."

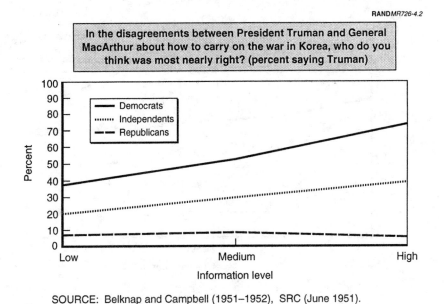

RAND*MR726-4.2*

SOURCE: Belknap and Campbell (1951–1952), SRC (June 1951).

Figure 4.2—Percentage Supporting Truman in Dispute with MacArthur

By November 1952, Republican support for escalation appears to have dwindled, and more knowledgable Republicans were willing to/ settle for a peaceful settlement. More knowledgeable independents, however, continued to reject this option. Figure 4.3 describes public attitudes in November 1952 toward Truman's preferred option of continuing the war until a peaceful settlement could be reached, broken down by partisan orientation and media usage.[25]

In this case, the data show that, when given a three-way choice between taking a stronger stand, pulling out, and continuing to try to get a peaceful settlement, the low-information Republicans (those least likely to be familiar with the party position) were most like

---

[25]There was no 1950 NES. The figure was constructed from raw data from the November 1952 NES found in National Election Studies and Inter-University Consortium for Political and Social Research (1995).

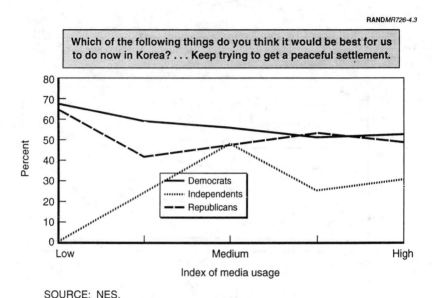

SOURCE: NES.

**Figure 4.3—Support for Trying to Get a Peaceful Settlement,
November 1952**

Democrats in supporting a peaceful settlement. At the second information level, the gap between Democrats and Republicans is greatest, about 17 points, and narrows thereafter. Only at the highest information level do the Democrats and Republicans polarize again in a predictable way. It is noteworthy that higher-information independents seemed least likely to support this option, possibly because they were following General MacArthur's position most closely.[26]

Figure 4.4 presents data on support for the escalatory option that MacArthur and many Republican leaders preferred, by party and information level. The figure shows that, as the information level of respondents increased, support for taking a stronger stand increased

---

[26]The zero value for the low-information independents may be due to the sparseness of the data, since cross-tabulation of these data by party and information level resulted in a very small number of observations in many cells.

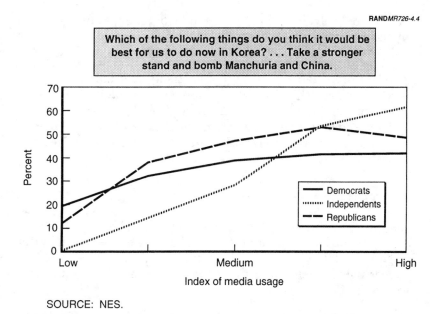

RANDMR726-4.4

Figure 4.4—Support for Taking a Stronger Stand in Korea,
November 1952

not just for Republicans, but for Democrats and independents as well.[27] Nevertheless, only for those Republicans at the second-to-highest information level did an actual *majority* of Republicans support this option, and at only two information levels did a majority of independents support this option. The overall result was that in none of the three groups did a majority support taking a stronger stand.[28] At the highest information level, however, independents

---

[27]The dramatic difference in willingness to take a stronger stand between low- and high-information Republicans may reflect divisions between the isolationist and internationalist wings of the Republican party. For isolationists, Korea was unimportant Asian real estate and not worthy of additional sacrifices.

[28]Overall, 38 percent of the Democrats, 41 percent of the independents, and 47 percent of the Republicans supported this option.

were more likely than Republicans to favor taking a stronger stand, and Democrats the least likely.[29]

In short, the evidence shows that support and policy preferences followed predictably from party leadership and individual differences in partisanship and information levels: Political leaders seriously considered only two options, and public support for these options tended to follow party preferences closely.[30] As shown in the previous chapter, however, escalation sentiment never reached overwhelming levels of support.

VIETNAM

Like Korea, Vietnam also provides evidence of the importance of leadership in declining support and policy preferences. A significant reason for the decline is to be found in the initially muted but increasingly strident criticism of the war by members of Congress and other opinion leaders.

The Fulbright hearings in early 1966 had been highly critical of the war and had featured establishment stalwarts criticizing the war, thus making dissent respectable. Over time, congressional restiveness grew as a result of the mounting costs and the lack of progress. Growing congressional and elite criticism of the war called into question both the importance of U.S. interests in Vietnam and the wisdom of incurring further costs in an intervention that seemed to be yielding little evidence of progress. By the Tet offensive in early 1968, vocal congressional opposition to the war was both common and widely reported by the media.[31]

---

[29]The high support from independents suggests that they may have been following General MacArthur on the matter.

[30]See Belknap and Campbell (1951–1952) on this point.

[31]Zaller (1992), Hallin (1986). Mueller (1973), pp. 116–117, notes that partisan differences in leadership support had become ambiguous by the second year of the war. Zellman (1971) traces the growth of senatorial opposition during the Vietnam War and suggests that public dissent increased markedly in 1965 and 1966; the rolls of opponents in the Senate grew from about 15 dissenters by March 1966 (eight of whom were quite vocal in their dissent as early as 1965) to about 25 by March 1968.

Support for the War

Public support for the Vietnam War evidences rather consistent partisan differences.[32] As long as it was President Johnson's war, public support for the war was generally higher among Democrats than Republicans. When it became a Republican war, Republicans were somewhat more inclined to express support for the war than were Democrats.[33]

One important difference between the Korean and Vietnam wars is that the most visible leadership opposition to Vietnam came from within the Democratic president's own party, including Democratic members of Congress. The leading role of this Democratic dissent can be inferred from the fact that mainstream Democratic politicians began to turn against the war at the high tide of popular support for the war in 1966.[34] However, the average Democrat in the public did not shift away from support of the war until late 1967 or early 1968.[35]

Equally important was the growing polarization between liberal and conservative leaders, which led in a predictable way to polarization within the public, with the gap the widest between the most knowledgeable in each group. Figure 4.5 plots support for the war from liberals and conservatives at different levels of political sophistication for 1966 (the solid lines) and 1970 (the dashed lines).

The figure shows that support for the war from both groups generally declined from 1966 to 1970, but that there are important ideological differences. The greatest decline was for the most politically sophisticated liberals, while the smallest decline was for the most sophisticated conservatives. In short, polarization was greatest between

---

[32]See Mueller (1973), pp. 115–154, for an analysis of the partisan and ideological correlates of support for the war. Mueller shows that there were differences in both the level and kind of support that paralleled differences among partisan leaders.

[33]See Mueller (1973), Table A.1, for data.

[34]Zaller (1992), p. 270. It is difficult to make the case, for example, that Senator Fulbright and Representative Tip O'Neill—early critics—were responding to constituency pressure.

[35]Mueller (1973), Table A.1, presents AIPO data showing that, in July 1967, 55 percent of Democrats supported the war. By October 1967, support had fallen to 48 percent and, by early February 1968, to 45 percent.

those who would have been most familiar with their opinion leaders'/
positions—high-information liberals and conservatives.

Policy Preferences

Vietnam also provides another case in which the most knowledgeable members of both parties—those who were most likely to be
aware of and embrace the positions of their party leaders—were
most like their leaders in polarizing on the question of war policies./
Figure 4.6 presents data on the policy preferences over time of the
most politically knowledgeable Democratic and Republican respondents in the NES surveys from 1964 to 1970.[36]

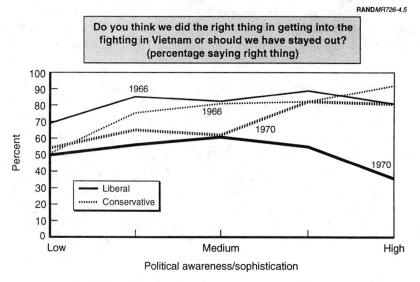

Figure 4.5—Changing Support for the Vietnam War, 1966–1970

---

[36]An index of political knowledge was constructed from three factual questions. The
index merely counts the number of correct responses, from zero to three. This is
somewhat similar to the approach Zaller (1987 and 1992) used.

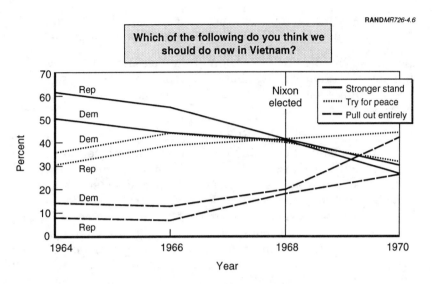

SOURCE: NES.

**Figure 4.6—Preferred Policies in Vietnam, High-
Knowledge Respondents**

The figure shows that sentiment for taking a stronger stand (the solid
lines) among the most politically knowledgeable in both parties
declined after 1964. By contrast, sentiment for pulling out entirely
increased among both Republicans and Democrats, although by
1970 high-knowledge Democrats were far more likely to support this
option than were Republicans. Finally, support for continuing the
war until a peaceful settlement could be reached was the preferred
option of knowledgeable Republicans in 1970, while the most knowledgeable Democrats preferred withdrawal to continuation of what
was by then a Republican war.

These data again show that those most likely to be aware of and
responsive to their party leaders' preferences on the war were also
the most inclined to share these preferences.[37] In short, Vietnam

---

[37]As Mueller (1973, p. 101) summarized the process after 1968:

provides another example in which widening divisions among party leaders appear to have led to growing divisions in the public.

Before concluding this discussion of the Vietnam War, a few additional words are necessary about the respective roles played by liberal opinion leaders who opposed the war and about the antiwar movement.

Before 1966, opposition to the expansion of the U.S. role in Vietnam was predominantly liberal and was well-represented not only within the early antiwar movement but also within both the Kennedy and Johnson administrations.[38] Liberal members of the political establishment who early on opposed the war almost surely interacted with intellectuals and other early members of the antiwar movement. The antiwar movement subsequently became radicalized and distinctly anti-establishment in character, however, which made potential opponents of the war in the political establishment more careful about being associated with the movement, probably stiffened the resolve of the average American, and may ultimately have prolonged the war.[39] As was described above, the avant-garde in the liberal establishment, including many senior members of the Johnson administration, continued to turn away from the war and came to favor de-escalation and withdrawal.[40] In short, while establishment liberals shared with demonstrators a preference for de-escalation, growing opposition to the war from respected members of the establishment clearly had a greater impact on increasing support for de-escalation in the mass public than did the antiwar movement.[41]

[With] the ultimate acceptance in mid-1969 by the Nixon Administration of a policy of gradual withdrawal while it also continued and formalized the policy of Vietnamization . . . withdrawal of a sort became official presidential policy and administration followers could move to its support. At the same time, of course, leaders of the Democratic opposition became increasingly withdrawal conscious, and people inclined to follow *their* lead had a model.

[38]Garfinkle (1964), p. 507.

[39]See Mueller (1973), p. 164 on this point.

[40]See Lorell et al. (1985), especially pp. 61–85.

[41]There is the possibility, however, that some came to oppose the war as a result of the very divisive domestic situation, i.e., that they did not believe the war was worth tearing the country apart.

THE GULF WAR

The Gulf War enjoyed very high levels of support at its start and was concluded well before the initial rally in support could decay. It furthermore achieved a decisive outcome at very low cost, resulting in very high levels of retrospective support. The Gulf War is also a good example of a president managing to maintain his leadership of the debate in spite of opposition from congressional opponents.[42]

President Bush benefited from bipartisan congressional support for his initial defensive measures in August 1990 and from high levels of public support. There is, however, good reason to believe both that U.S. political leaders could have become divided over the war had it not turned out so well and that these divisions would have been paralleled in the public.[43]

For example, President Bush's announcement of an increase in forces to create an "offensive option" in November 1990 was met with congressional hearings and harsh criticism from opponents, mostly Democrats. The consequence was that the public's approval of the president's handling of the Gulf crisis appears to have declined nearly ten points.[44] Support for sending the additional troops was partisan.[45] Leadership support for continued reliance on sanctions and the congressional vote to authorize the use of force in the Gulf also split largely along partisan lines.

Even the rally in support at the beginning of the war had a strong partisan component, conservatives being far more likely to follow conservative leaders in supporting the war.[46] Zaller (1992) examined

---

[42]Zaller (1992) described the mobilization of support in August 1990 as "one of the most striking cases of elite opinion leadership examined in this book." Zaller (1992), p. 269.

[43]Mueller (1994) also makes this argument.

[44]See Mueller (1994), Tables 8–14.

[45]Zaller (1992), p. 104. Zaller also shows that, when President Bush was more conciliatory, support for tougher action fell among both Republicans and Democrats, but when he was more threatening, support increased among Republicans and declined among Democrats. Zaller (1992), p. 107.

[46]According to data in Mueller (1973), p. 271, the rally at the beginning of the Vietnam War seems also to have had a partisan component. The change in support from May

panel data from the University of Michigan's 1990 NES and discovered that the rally at the beginning of the Gulf War had a strong partisan component to it, with the greatest rally being found among the most knowledgeable conservatives and the smallest rally being found among the most knowledgeable liberals (see Figure 4.7).

The "before" lines in the top two panels of the figure represent support for going to war in the November 1990 survey, while the "after" lines are associated with the January 1991 follow-up survey. The "probability of attitude change" in the bottom panel of the figure summarizes the probability of rallying to support the war for conservatives and liberals at various knowledge levels. In short, the run-up to the Gulf War provides another example in which divisions among the public followed predictably from divisions among their leaders.

The data also strongly suggest that, if the war had gone badly and political leaders had polarized on the basis of ideology or partisanship, this would have widened the fissures along partisan or ideological lines in the public.[47] Even if the war had gone worse than expected, however, the U.S. experiences in Korea and Vietnam lead us to believe that President Bush would have had some period of time—probably in the range of several months to perhaps as long as a year—to conclude the conflict on favorable terms before mainstream political leaders turned against the war. In light of the formidable military capabilities of the United States, it is somewhat difficult for this author to imagine the conflict lasting this long.[48]

1965 (before President Johnson's escalation) to November 1965 was seven points for Democrats and independents, but only four points for Republicans.

[47]Indeed, congressional leaders had already divided along partisan lines before the war, and the beginnings of a public antiwar movement could be seen in the fall of 1990. Despite the fact that Democratic party leaders made the vote one of conscience rather than party loyalty, the congressional vote to authorize the use of force was nearly a straight party-line vote, with Republicans supporting and Democrats opposing. John Mueller describes this as a "duel" between President Bush and his supporters, on the one hand, and his opponents, on the other. See Mueller (1994), pp. 58–60.

[48]Mueller (1994), for example, notes that strategies were available to Iraq (e.g., an Alamo-like stand that the Iraqis could have made in Kuwait City) that could have prolonged or increased the costs of the conflict. Nevertheless, counterstrategies were also available to the U.S. coalition. In short, we are in the realm of a counterfactual world on this issue.

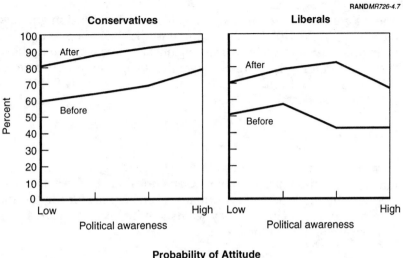

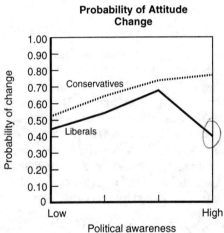

SOURCE: Zaller (1993).

Figure 4.7—Probability of Rallying to Support the Gulf War

And if Korea and Vietnam are good guides to how events might have played out in the Gulf War, even if the war had gone on longer and mainstream politicians had turned against it, there would have been little support for quitting the intervention before an orderly conclusion—most importantly including the safe return of U.S. POWs—could be achieved.

SOMALIA

Somalia is an interesting case in which bipartisan support for the initial intervention turned to bipartisan opposition by the early fall of 1993. Rather than being partisan, however, support for the Clinton administration's Somalia policy seems to have been associated with positive evaluations of the president.

Support

Political leaders gave strong bipartisan support to the U.S. intervention in Somalia. This support seems generally to have held until the summer of 1993, by which time the initial objectives had been achieved and the mission had changed. By September 1993, congressional opposition to the operation in Somalia had also become bipartisan. Both houses of Congress had passed a nonbinding resolution calling on the Clinton administration to seek approval by October 15 for keeping U.S. forces in Somalia and threatening a cutoff in funds if such action was not taken.[49] Public support had also declined to about four in ten by this time.

With the deaths in Mogadishu in early October 1993, members of Congress on both sides of the aisle opposed continuation of the operation, although the president ultimately prevailed upon the Congress to approve an orderly withdrawal by the end of March 1994. By this time, there were only minor partisan differences among the public in evaluations of Somalia policy, with only about one in three supporting.

---

[49]Hirsch and Oakley (1995), p. 127.

Although support for the president may be somewhat conflated with support for his policies, hard-core support for the president seems to have been more important than partisanship in evaluations of President Clinton's handling of Somalia in October 1993. Thirty-eight percent of Democrats (as opposed to 25 percent of Republicans) approved of the president's handling of Somalia, while 50 percent of those who approved of the president's job handling (as opposed to 13 percent who did not) and 65 percent who approved of his handling of foreign affairs (as opposed to 28 percent who did not) approved of his handling of the situation in Somalia.[50]

Policy Preferences

Support for the president, not partisanship, may also have been associated with policy preferences. According to Gallup's October 8–10 poll, only 33 percent of Democrats and 32 percent of Republicans wanted to complete the humanitarian mission in Somalia.[51] Larger differences emerge when the contrast is by evaluations of presidential performance: only 35 percent of those who approved of the president's job handling supported keeping U.S. troops in Somalia until the humanitarian mission was accomplished (as opposed to 26 percent who did not), and 43 percent who approved of his foreign-affairs handling supported keeping troops in Somalia until the mission could be accomplished (as opposed to 23 percent who did not approve).[52]

In the case of Somalia, then, bipartisan support from leaders turned to bipartisan opposition, and this decline in support was closely associated with a parallel decline in public support. In the end, the president's supporters were following his lead, while opponents were following his congressional opponents.

---

[50]Gallup, October 8–10, 1993.

[51]The corresponding percentages supporting withdrawal were 65 and 62 percent.

[52]The corresponding percentages supporting withdrawal were 62 and 56 percent.

CHAPTER CONCLUSIONS

This chapter has provided substantial evidence supporting the propositions that leadership consensus or dissensus is an essential element in the character of public support for U.S. military interventions and that leadership divisions tend to cue divisions among the public in a predictable way. In short, when there is bipartisan consensus among leaders in support of an intervention, divisions within the public are generally muted; when there are partisan divisions among the leaders, the public tends to become divided along the same lines.

The analysis also suggests that beliefs about benefits, prospects, acceptable costs, support, and policy preferences can differ across partisan or ideological groups, leading to different levels of support and policy preferences.[53] In the Korean War, for example, isolationist Republicans did not consider Korea to be important enough to be worth the lives of U.S. servicemen, and they were accordingly less likely to prefer escalation. Internationalist Republicans, on the other hand, believed it important enough to widen the war to include attacks on Manchuria, even at the risk of a larger war. Most Democrats and independents thought the stakes important enough to reject withdrawal and continue the war until peace could be achieved but not important enough to risk a wider war. In Vietnam, growing polarization among leaders affected support and policy preferences in a similar way. In Somalia, by contrast, bipartisan leadership support turned to bipartisan opposition and a desire for an orderly withdrawal, and these preferences were mirrored in the public.

The stylized argument presented in this chapter suggests that support—and the evaluation of benefits, prospects, and costs that was described in the last chapters—is socially constructed. The media report debates among leaders and experts to members of the public, who consider and discuss them. The media subsequently poll these same members of the public, informing leaders of the success of their persuasive arguments. While something of a simplification, this

---

[53]In Kagay's (1992) words "leadership and events matter."

characterization captures some of the most important features of how the democratic conversation works.

The next chapter will draw together the various threads of this analysis and provide conclusions.

Chapter Five

CONCLUSIONS

When asked to support a military operation, the American public ultimately must weigh the intangible benefits of achieving foreign policy objectives against the most tangible costs imaginable—the lives of U.S. service personnel. The metaphor of an ends-means calculus can be used to understand the factors that are associated with support for military operations and a willingness to ask others to sacrifice their lives. This metaphor characterizes support as being the result of a series of tests or questions that political leaders and the public answer collectively:

- Do the benefits seem to be great enough?

- Are the prospects for success good enough?

- Are the expected or actual costs low enough?

- Taken together, does the probable outcome seem (or still seem) to be worth the costs?

Assessed in light of these questions, the historical record suggests that the role of casualties in domestic support for U.S. wars and military operations is somewhat different from the conventional wisdom.

When we take into account the perceived benefits of the operation, broadly conceived as the importance of the interests at stake and the principles being promoted, the evidence of a recent decline in the willingness of the public to tolerate casualties appears rather thin. The historical record in fact suggests a rather high degree of differentiation in the public's willingness to tolerate casualties, based upon the merits of each case.

Whenever the reasons for introducing U.S. forces lack either moral force or broadly recognized national interests, support may be very thin indeed, and even small numbers of casualties may often be sufficient to erode public support for the intervention. For in the end, most Americans do not want lives to be sacrificed for any but the most compelling and promising causes, and they look to their leaders to illuminate just how compelling and promising the causes are.

The Gulf War, for example, was a recent military operation in which a majority viewed important principles and interests to be at stake and showed a rather higher willingness to tolerate casualties than most realize, in many ways much closer to Korea and Vietnam than other cases. By the same token, the unwillingness of the public to tolerate very high casualties in some recent U.S. military operations has had to do with the fact that majorities—and their leaders—did not perceive the interests and principles at stake to be particularly important.

When they approve of a military operation, members of the public typically grant the president wide latitude to conclude the operation in the fashion he chooses. This permissive environment can be lost, however, if the operation does not live up to the expectations upon which initial support was premised. This often leads to polarization among leaders (and within the public) over the best policy for concluding the operation. Contrary to the conventional wisdom, increasing casualties and declining support do not necessarily lead to majority support for immediate withdrawal. Contrary to the counter-conventional wisdom, casualties do not necessarily lead to majority demands for "escalation to victory."

Instead, the preferences of individual members of the public—whether escalatory or de-escalatory—seem most closely associated with an assessment of the U.S. equities in the situation and the credibility of the alternatives that leaders and experts offer. Credibility is often judged on the basis of partisan or ideological cues. In Korea and Vietnam, despite the polarization and some support for the extreme options of immediate withdrawal and escalation of the war, the ultimate result was a grudging willingness to continue each war until an orderly withdrawal—including the return of U.S. prisoners of war—could be accomplished. In Somalia, a majority also preferred an orderly withdrawal following the return of U.S. servicemen that Aidid held hostage and rejected both immediate withdrawal and an increased—or extended—commitment. For most, the U.S. equities did not justify additional efforts to save Somalia from itself.

As a result of the Gulf War, the public does not expect—and is unlikely to demand—that all future U.S. military operations be bloodless. Indeed, it is more accurate to say that the public hopes for low-to-no casualty operations but fears a very different outcome. A majority of the public will accordingly continue to support a range of measures to minimize American casualties in wars and military operations: diplomacy to foster a more benign environment for U.S. forces; cost- and risk-sharing with allies; strategies, tactics, doctrine, and training; and force structures and technologies that can minimize U.S. casualties. Nevertheless, the linkage of these pieces is not very well understood.

In an era of limited resourcing for defense, the implications for strategy, research and development, and force planning are also not particularly well understood. The historical record suggests that a majority of the American public will be more willing to accept casualties when important interests and principles are at stake—most likely including the current major planning contingencies centered on Iraqi and North Korean aggression—and is least willing to accept losses in the sorts of operations that the nation has most recently undertaken—armed interventions in failed states. Put simply, is the ability to conduct a low-to-no casualty peace operation in Somalia, Haiti, or Bosnia more desirable than a similar capability for a war in the Gulf or Northeast Asia, where casualties could easily be much higher?

IMPLICATIONS FOR POLICYMAKERS

One of the key findings of this research is the central role of leadership in determining domestic support for U.S. military involvements. The calculus described above masks a much richer social process linking public support or disaffection to leadership consensus or conflict, although leadership needs to be viewed in a broader sense than it is typically conceived.

Large segments of the public rely upon political leaders to vet the often complex issues involved in prospective and ongoing military interventions. These groups respond predictably when the leaders they find most credible begin to question—or decide to oppose—an intervention. When support or the preferred strategies for conclud- ing the operation fall prey to partisan divisions among leaders, the public will typically also become divided. In short, when political and other opinion leaders fail to agree with the president that much (or any) good is likely to come of an intervention, there should be little surprise that the public also becomes divided.

There have been many disagreements among leaders about whether the merits of recent U.S. military actions in the Gulf War, Somalia, Haiti, and Bosnia have justified their possible costs. While it is not entirely unhealthy for a democracy to weigh carefully its decisions on the use of force, the potential consequences of these recurring dis- agreements among leaders are quite sobering. They can lead to enduring divisions in the public and to support that is brittle and easy for adversaries to exploit, thereby leading both to failed inter- ventions and incorrect lessons for the future. Ultimately, such dis- agreements may erode the credibility of threats of force to protect important U.S. interests. The irony, of course, is that when deter- rence and coercive diplomacy fail, the costs to the nation may turn out to be even higher.

Policymakers who are mindful of the premises under which support has been given for a particular U.S. military operation will often be able to build and sustain a permissive environment for conclusion of the operation. They are also most likely to understand the con- straints on—and opportunities for—presidential leadership when dramatic change occurs and initial support has eroded. But until U.S. leaders arrive at a new bipartisan consensus on the role of mili- tary force in the post–Cold War world, we should expect disagree- ments among leaders whenever the U.S. deploys its forces, and these disagreements will continue to foster divisions among the public. The absence of a larger foreign policy consensus will contribute to support that is often shallow and highly responsive to the costs in

casualties. As the historical record shows, however, attributing ✓✓
declining support solely to casualties misses the real story.

PUBLIC OPINION DATA

A few words about the use of polling data are in order, the most important being that public opinion data can be exceedingly treacherous, and many analyses and news stories fail to take this sufficiently into account.

First, there are problems with the availability of data, often making comparisons over time or across cases exceedingly difficult. Analysts of public opinion on U.S. military interventions often do not have public opinion data for the questions that would have been most useful; instead, they must rely upon the questions that media organizations and scholars asked at the time. Instead of having the panel data (time series data based upon asking the same respondents questions over time), which would enable the analyst to assess the attitudes of the same respondents (thereby being able to gauge the level of instability in attitudes), the analyst often lacks either consistently worded time series or breakdowns of attitudes among subgroups. The result is that public opinion analysis is exceedingly tedious and detailed work, fraught with opportunities to misread the data.

Second, there is disagreement about what public opinion data actually measure. Questions are often asked of respondents who have never given very much thought to the matters about which they are being asked and who certainly do not have any fixed opinion on the subject. Responses are most often top-of-the-head answers to questions and do not reflect deeply held beliefs. At the level of the respondent, there is often a great deal of response instability, with

individuals changing responses for no apparent reason.[1] To compli-
cate matters further, even slight changes in question structure or
wording can affect responses when attitudes are uncrystallized, and
there is no easy way to determine the level of crystallization other
than examining as much as possible of the available data.

On the bright side, instability in individual responses tends to cancel
itself out in the process of aggregation, leaving reasonably reliable
population estimates.[2] Seemingly contradictory polling results, furthermore, often turn out to be consistent and reasonable when one
takes into account critical differences in wording or when interpretation is informed by still other public opinion data or identification
of critical events. Responses among subgroups of the population can
also vary, often providing deeper insights into the correlates of particular attitudes, and suggesting a more coherent patterning of
beliefs than is found at the aggregate level. Differences in partisanship or ideology, or in political knowledge or sophistication, for
example, can lead to an understanding of important intergroup differences. And by taking into account *both* partisanship or ideology
and political knowledge and sophistication, important subgroup
differences can often be understood rather well.

Table A.1 provides data for Figure 2.1, which plotted support for a
generic military operation against hypothetical losses of life. Figure
2.1 plots the cumulative percentages. For example, 93 percent (100
percent minus the 7 percent who refused to answer or did not know)
would find no casualties an acceptable number. Similarly, 81 percent
(the 93 percent minus the 12 percent who would only accept no
casualties) would find one death an acceptable number.

What follows describes the data that were used in Figure 2.2, which
plots support as a function of casualties in a number of U.S. military
operations. The reader will note that the wording of the questions
varies; the data presented are, however, reasonably representative of
the available data.

---

[1]This problem is compounded when there are in fact plausible explanations (e.g.,
changing conditions or events) of which the analysts themselves are blissfully
unaware.

[2]Page and Shapiro (1992).

Table A.1

Support as a Function of U.S. Battle Deaths

> I would like to get some idea of what you think "too much loss of life" is in a military intervention. What would be the rough figure you would use as an acceptable number of U.S. deaths?

Number	Support (percent)	Cumulative (percent)
None	12	93
One	22	81
Ten	9	59
100	16	50
1,000	16	34
10,000	10	18
100,000	3	8
1,000,000	1	5
>1,000,000	4	4
Refused to answer	5	—
Did not know	2	—

SOURCE: Americans Talk Issues (June 23–July 1, 1991).

NOTE: Cumulative percentages identify that number or a higher number of casualties as "an acceptable number of U.S. deaths."

THE SECOND WORLD WAR

Data for the Second World War (Table A.2) proved to be problematic because of the absence of trends made up of consistently worded questions, and the lack of availability of accurate data on battle deaths at the time of each poll. This case was included, however, to provide a basis for comparison of the other cases.

Public opinion data on support for the war were taken from Campbell and Cain (1965). Estimates of the total cumulative battle deaths over time were constructed from Clodfelter's (1992) estimates of the number of battle deaths in the various campaigns of the Second World War. These estimates, while crude, would be expected

Table A.2

Support for World War II as a Function of
U.S. Battle Deaths

Source	Date	Estimated Deaths	Percent Supporting	Wording
AIPO	December 1941	2,400	86.0	A
AIPO	February 1942	16,247	87.0	A
AIPO	February 1943	17,899	92.0	A
AIPO	September 4, 1943	24,127	89.0	A
AIPO	February 2, 1944	30,736	77.0	B1
AIPO	April 10, 1946	217,000	77.0	B2
Fortune	March 1943	20,737	85.5	C
NORC	May 7, 1944	36,274	81.0	D
OPOR	August 30, 1944	67,806	82.0	E
AIPO	February 19, 1945	166,183	75.0	F
Fortune	June 1, 1945	212,543	86.8	G

The percentages reflect responses that connote support for or a desire to continue the war:

A. Want peace as things are now? No.

B1. Do you think you, yourself, will feel it was a mistake for us to have entered this war? No.

B2. Do you think it was a mistake for the United States to enter World War II? No.

C. Keep on fighting? Yes.

D. Demand unconditional surrender before stop fighting Germany? Yes.

E. Discuss peace with Hitler now? No.

F. Approve of requiring unconditional surrender? Yes.

G. Go on and clean out Japs in China? Yes.

SOURCES: Campbell and Cain (1965); AIPO; *Fortune*; NORC; OPOR.

to approximate roughly—but underestimate somewhat—the level of casualties in any given month of the war.[3]

THE KOREAN WAR

Public opinion data for Korea (Table A.3) were taken from Mueller (1973, Table 3.1). Casualty data are from the Korean Conflict Casualty File (KCCF), created by the Directorate of Information Operations and Reports of the Washington Headquarters Services, Office of the Secretary of Defense, and maintained by the National Archives. The "Supporting" column reflects those who responded "no" to the question.

Table A.3

Support for Korea as a Function of U.S. Battle Deaths

Do you think the United States made a mistake in going into the war in Korea, or not?		
Date of Poll	Cumulative Hostile Deaths	Supporting (percent)
August 1950	4,631	66
December 1950	13,991	39
February 1951	16,716	41
March 1951	17,602	43
April 1951	18,674	45
June 1951	20,641	42
Early August 1951	21,459	47
March 1952	25,617	37
September 1952	28,185	39
October 1952	29,202	36
Late October 1952	29,874	37

SOURCE: AIPO.

---

[3]There were, for example, a total of about 294,000 battle deaths in the war; the total casualties from Clodfelter's estimates of campaign-related battle deaths is 225,000.

THE DOMINICAN REPUBLIC

There appeared to be only a couple of questions that asked about support for the Dominican intervention (Table A.4). Casualties were estimated from Clodfelter (1992).

THE VIETNAM WAR

Public opinion data for the Vietnam War (Table A.5) were taken from Mueller (1973), Table 3.3. Casualty data are from the Combat Area Casualties Current File (CACCF) as of October 1994, created by the Directorate of Information Operations and Reports of the

Table A.4

Support for Dominican Intervention as a Function of U.S. Battle Deaths

How do you feel about President Johnson's sending troops into the Dominican Republic (1965)?	
	Responding (percent)
Approve	76
Disapprove	17
No opinion	7

SOURCE: Gallup/GOI (May 13–18, 1965), at 1 battle death.

As you probably remember, in late April of this year, the United States sent troops into Santo Domingo. Do you think the United States did the right thing or the wrong thing in deciding to send troops into Santo Domingo?	
	Responding (percent)
Right thing	52
Wrong thing	21
No opinion	27

SOURCE: Gallup/GOI (November 18–23, 1965), at 28 battle deaths.

Table A.5

Support for Vietnam as a Function of U.S. Battle Deaths

Date of Poll	Cumulative Hostile Deaths	Supporting (percent)	Wording
August 1965	166	61	A
November 1965	924	64	B
March 1966	2,415	59	A
May 1966	3,191	49	A
September 1966	4,976	48	A
November 1966	5,798	51	A
February 1967	7,419	52	A
May 1967	10,341	50	A
July 1967	11,939	48	A
October 1967	13,999	44	A
December 1967	15,695	46	A
February 1968	19,107	42	A
March 1968	20,658	41	A
April 1968	22,061	40	A
August 1968	27,280	35	A
October 1968	28,860	37	A
February 1969	32,234	39	A
September 1969	38,581	32	A
January 1970	40,112	33	A
March 1970	40,921	32	A
April 1970	41,479	34	A
May 1970	42,213	36	A
January 1971	44,109	31	A
May 1971	44,980	28	A

A. In view of the developments since we entered the fighting in Vietnam, do you think the U.S. made a mistake sending troops to fight in Vietnam?

B. Some people think we should not have become involved with our military forces in Southeast Asia, while others think we should have. What is your opinion?

SOURCE: AIPO.

Washington Headquarters Services, Office of the Secretary of Defense, and maintained by the National Archives. Those who responded "no" to the first question and those who responded that the United States should have become involved are counted as supporters.

LEBANON

The data for Lebanon (Table A.6) were based upon a "mistake" question. Cumulative casualties at various times were estimated from a variety of sources, including Clodfelter (1992), the Department of Defense, *Facts on File*, and the *New York Times Index*.

PANAMA

Public opinion data for Panama are from the questions reported in Table A.7. Casualty data were from Clodfelter (1992), the Department of Defense, and from "Statement by Press Secretary Fitzwater on United States Military Action in Panama, December 21, 1989" (1990, pp. 1726–1727), which reported that there had been 18 battle deaths as of December 21.

THE GULF WAR

Data for the curve representing the Gulf War are from the public opinion question asked by the *Los Angeles Times* on November 14, 1990 and reported in Table A.8. This result is broadly representative of the prewar sentiment regarding approval and casualties.

Table A.6

Support for Lebanon as a Function of U.S. Battle Deaths

Do you think the United States made a mistake in sending the Marines to Lebanon, or not?		
Dates of Poll	Hostile Deaths	Supporting (percent)
October 7–10, 1983	12	37
November 18–21, 1983	253	45
December 9–12, 1983	253	44
January 13–16, 1984	257	39
February 10–13, 1984	266	33
February 10–12, 1984	266	34

SOURCE: Gallup.

Table A.7

Support for Panama as a Function of U.S. Battle Deaths

Source	Date(s) of Poll	Hostile Deaths	Supporting (percent)	Wording
ABC News	December 20, 1989	0	80	A
USA Today	December 20, 1989	0	81	B
ABC News	December 21, 1989	18	80	A
Gallup/*Newsweek*	December 21, 1989	18	80	C
ABC/*Washington Post*	January 11–16, 1990	23	82	A

Wordings for the questions were as follows:
A. Do you approve or disapprove of the U.S. having sent its military forces into Panama to overthrow (Manuel) Noriega?
B. Tuesday night (12/19/89) President Bush sent U.S. military forces into combat in Panama to oust military leader Manuel Noriega. Do you approve or disapprove of President Bush's decision to send troops to Panama?
C. Was the U.S. justified or not in sending military forces to invade Panama and overthrow (General Manuel) Noriega?

These data were converted to the data in Table A.9, which was the basis for the curve for the Gulf War in Figures S.1, 2.2, and 2.7. First, I assumed that all of the 77 percent who said they thought it was worth risking American lives would support the war if there were in fact no casualties, i.e., the curve intersected the y-axis at 77 percent. Next, to facilitate the drawing of the curve, the "less than 1,000" option was recoded to 500 and the "more than 20,000" option was recoded to 50,000. Finally, cumulative percentages were computed directly from the percentages in the table to give the percentage who were willing to accept that or a higher level of casualties. For example, 25 percent thought it worth fighting a war if there were more than 20,000 U.S. deaths; 36 percent (25 plus 11) thought it worth fighting at 20,000 or more; and so on. Final percentages were then computed by multiplying by 0.77, the percentage who initially said it was worth risking American lives to protect our oil supplies and/or demonstrate that countries should not get away with aggression.[4]

---

[4]Since there are other possible reasons that respondents would have thought the war worth risking lives (e.g., to eliminate nuclear weapons, to stop the atrocities in Kuwait,

Table A.8

Support for Gulf War as a Function of U.S. Battle Deaths

	Agree (percent)
<1,000 (recoded to 500 in figure)	15
1,000	12
5,000	10
10,000	7
20,000	11
>20,000 (recoded to 50,000 in figure)	25
Not sure	15
Refused to answer	5

You just said that it's worth risking American lives (to protect our oil supplies and/or to demonstrate that countries should not get away with aggression.)
—Do you think it's worth fighting a war with Iraq if 1,000 American soldiers are killed?
—(If says 1,000) Do you think it's worth fighting a war if 5,000 American soldiers are killed?
—(If says 5,000) Do you think it's worth fighting a war with Iraq if 10,000 American soldiers are killed?
—(If says 10,000) Do you think it's worth fighting a war with Iraq if 20,000 American soldiers are killed?
(Accept less than 1,000 killed and more than 20,000 killed as volunteered responses)

SOURCE: *Los Angeles Times* (November 14, 1990).

NOTE: Asked of those who said it is worth risking the lives of American soldiers to protect our oil supplies (29 percent) or to demonstrate that countries should not get away with aggression (48 percent, a total of 77 percent of those polled).

The data in Table A.8 reflect the recoding to capture the percentage who thought it worth risking American lives at each hypothetical casualty level.

---

to topple Saddam), this 77 percent may actually underestimate somewhat the percentage willing to take the risk of lost U.S. lives.

Table A.9

Support for Gulf War as a Function of
U.S. Battle Deaths, Converted

> You just said that it's worth risking American lives (to protect our oil supplies and/or to demonstrate that countries should not get away with aggression.) Do you think it's worth fighting a war with Iraq if X American soldiers are killed?

	Agree (percent)
No casualties (total believing worth risk)	77
<1,000 (recoded to 500)	65
1,000	56
5,000	49
10,000	43
20,000	35
>20,000 (recoded to 50,000)	15

SOURCE: *Los Angeles Times* (November 14, 1990).

NOTE: Asked of those who said it's worth risking the lives of American soldiers in order to protect our oil supplies (29 percent) or in order to demonstrate that countries should not get away with aggression (48 percent), a total of 77 percent of those polled.

As can be seen in Figure 2.7, this question was generally representative of the various prewar public opinion questions that asked about support given various hypothesized casualty levels in the war; indeed, of the available questions, it provides a rather conservative estimate. The following figure compares these questions; as can be seen, the question that was used in Figure 2.2 is a rather conservative estimate of the prospective willingness of the public to accept casualties.[5]

SOMALIA

Somalia also posed a challenge because the "mistake" questions that were asked about Somalia did not appear to capture adequately the

---

[5]The text of these other questions can be found in the tables in Mueller (1994).

actual levels of public support for the operation. Nevertheless, one of the reviewers of this report noted that he is wary of putting questions that ask how a president is handling a situation (the "handling" questions in Table A.10, below) together with questions that ask if it has been a mistake to send troops. I strongly agree that these are very different questions that often get at very different phenomena. Only after careful study should analysts gauge support on the basis of questions other than the "mistake" question or pool data from different sorts of questions.

The difficulty in the case of Somalia arises in large part from the fact that the mission of the operation changed from the initial one of providing security to enable humanitarian relief to a broader one of rebuilding Somalia's political society. Many seem to have viewed the original humanitarian mission as having been the right thing to do, but came to oppose the broader missions that were undertaken in the spring and summer of 1993. "Mistake" questions—asking whether respondents thought the original operation was the right thing to do—would, accordingly, have overestimated support while underestimating the growing disaffection for the sort of operation that Somalia actually became.

To illustrate this point, Figure A.1 plots three types of questions from Table A.10. The figure presents a composite series for the "mistake" question (asking respondents whether they thought the initial intervention had been the "right thing" to do) constructed from questions from CBS, two separate series (one from NBC and one from *Time*/CNN) for "presence" questions (asking whether the respondent approved of U.S. forces being in Somalia), and a composite series for "handling" questions (asking whether respondents approved of the president's handling of Somalia).[6]

As the figure shows, the "mistake" questions from CBS/*New York Times* and CBS fell from an initial level of about eight out of ten to about six out of ten. By comparison, judged by the "handing" and

---

[6]The last data point for the "handling" line is an average of the percentages of the 13 "handling" questions that were asked at 28 deaths due to hostile action. Thus, the points for "handling" questions are C1, D1, D2, and an average of D3 through D13, E1 and E2, from Table A.10.

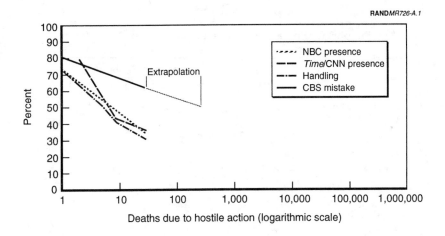

Figure A.1—Comparison of Questions on Somalia

"presence" questions support fell from an initial level that was roughly comparable—perhaps three out of four—to about one in three. That is, the "mistake" questions show a consistently higher level of support and fall at a slightly slower rate.

The inadequacy of the "mistake" question in describing support for Somalia becomes readily apparent when one imputes a straight-line extrapolation of the data for the "mistake" question: When extrapolated, it appears that support for Somalia as measured by the "mistake" questions would not have fallen to 50 percent until perhaps 250 casualties had been incurred in Somalia. As was described in the case study on Somalia, however, it is not only difficult to support the contention that six out of ten actually counted themselves as supporters of the operation following the deaths in Mogadishu, it is also quite doubtful that, at 250 battle deaths, 50 percent of the public would still have remained as supporters. Thus, the "mistake" question seems to overestimate support for the operation greatly.

By comparison, the only major differences between the "handling" and "presence" questions were that approval for NBC's "presence" questions started at a somewhat higher level than did either Time/CNN's "presence" questions or the series on presidential

Table A.10

Support for Somalia as a Function of U.S. Battle Deaths

Source	Date(s) of Poll	Hostile Deaths	Supporting (percent)	Wording
"Presence" questions				
NBC/*WSJ*	December 12–15, 1992	0	74	A1
Time/CNN	January 13–14, 1993	1	79	B1
Time/CNN	September 23–24, 1993	7	43	B2
NBC News	October 6–7, 1993	28	34	A2
Time/CNN	October 7, 1993	28	36	B3**
"Handling" questions				
CBS/*NYT*	December 7–9, 1992	0	73	C1
CBS/*NYT*	June 21–24, 1993	3	51	D1
Times Mirror	September 9–15, 1993	7	41	D2
ABC News	October 5, 1993	28	33	D3
NBC News	October 6, 1993	28	30	D4
CBS News	October 6, 1993	28	21	D5
CBS News	October 6–7, 1993	28	26	D6
ABC News	October 7, 1993	28	36	D7
WPOST	October 7–10, 1993	28	36	D8
Gallup	October 8–10, 1993	28	32	D9
ABC News	October 12, 1993	28	34	D10
CBS News	October 18–19, 1993	28	29	D11
Times Mirror	October 21–24, 1993	28	33	D12
ABC News	November 11–14, 1993	28	35	D13
Time/CNN	October 7, 1993	28	31	E1
Time/CNN	October 21, 1993	28	29	E2
"Mistake" questions				
CBS/*New York Times*	December 7–9, 1992	0	81	F
CBS News	October 6, 1993	29	64	G
CBS News	October 6–7, 1993	29	63	G
CBS News	October 18–19, 1993	29	67	G
CBS News	December 5–7, 1993	(29)	62	G

Table A.10—Continued

Wordings for the questions were as follows:

A. Should U.S. military forces be involved in the situation in Somalia, or should they not be involved? [Percent saying should be involved.]

B. In general, do you approve or disapprove of the presence of U.S. troops in Somalia? [B3** prefaced by: Here are a few questions concerning the recent events in Somalia, in which U.S. soldiers have been killed or taken prisoner by forces controlled by a Somalian warlord . . .] [Percent approving.]

C. Do you approve or disapprove of the way George Bush has been handling the situation in Somalia? [Percent approving.]

D. [Overall,] do you approve [or disapprove] of the way [President] Bill Clinton is handling the situation in Somalia? [Percent approving.]

E. Do you think President (Bill) Clinton is doing a good job [or a poor job]: [handling, dealing with] the situation in Somalia? [Percent saying "good job."]

F. Do you think the United States is doing the right thing to send U.S. troops to Somalia to try and make sure shipments of food get through to the people there, or should U.S. troops have stayed out? [Percent saying "right thing."]

G. Do you think the United States did the right thing to send U.S. troops to Somalia last December (1992) to try to make sure shipments of food got through to the people there, or should the U.S. troops have stayed out? [Percent saying "right thing."]

NOTE: Casualty data were from the Department of Defense.

handling of Somalia. Otherwise, all three series begin and end at roughly comparable levels. In short, the slight differences between the results for these two types of questions suggest that they can safely be pooled to estimate the slope of declining support as a function of casualties as was done here.

In short, a number of different series—asking about the president's handling of Somalia, whether the United States should be involved, and whether the presence of troops was approved—showed essentially the same result: Support for the operation started out at about three out of four and fell to about one in three. This seems to be a reasonably robust finding, consistent with data from the results of other of the more than 200 public-opinion questions that were asked contemporaneously about Somalia. The "mistake" question, on the other hand, while also showing declining support, seems to overestimate support for Somalia.

In many—perhaps most—cases, it is not a very good idea to base assessments of public support on questions on handling and pres-/ ence instead of those that ask whether the respondent feels that the intervention had been a mistake. In the case of Somalia, however, not only do they appear to provide a more accurate gauge of the actual levels of support that obtained than do the "mistake" questions, their comparability suggests that they can safely be pooled.

Ajzen, Icek, and Martin Fishbein, Understanding Attitudes and Predicting Social Behavior, Englewood Cliffs, N.J.: Prentice-Hall, Inc., 1980.

Belknap, George, and Angus Campbell, "Political Party Identification and Attitudes Toward Foreign Policy," *Public Opinion Quarterly*, Vol. 15, No. 4, Winter 1951–1952, pp. 601–623.

Braestrup, Peter, *The Big Story*, Boulder: Westview, 1977.

Brody, Richard A., "Public Evaluations and Expectations and the Future of the Presidency," in James Streling, ed., *Problems and Prospects for Presidential Leadership: The Decade Ahead*, Lanham, Md.: University Press of America, 1983.

_____, *Assessing the President*, Stanford, Calif.: Stanford University Press, 1991.

Brown, Seyom, *New Forces in World Politics*, Washington, D.C.: Brookings, 1974.

Burbach, David T., *Presidential Approval and the Use of Force*, working paper, MIT Center for International Affairs, Defense and Arms Control Studies Program, May 1994.

Campbell, Joel T., and Leila S. Cain, "Public Opinion and the Outbreak of War," *Journal of Conflict Resolution*, Vol. 9, September 1965, pp. 318–329.

Cantril, Hadley, *Gauging Public Opinion*, Princeton: Princeton University Press, 1947.

Cantril, Hadley, ed., with Mildred Strunk, *Public Opinion, 1935–1946*, Princeton: Princeton University Press, 1951.

Chong, Dennis, Herbert McClosky, and John Zaller, "Social Learning and the Acquisition of Political Norms," in Herbert McClosky and John Zaller, eds., *The American Ethos*, Cambridge, Mass.: Harvard University Press, 1984.

Clodfelter, Michael, *Warfare and Armed Conflicts: A Statistical Reference to Casualty and Other Figures, 1618–1991*, Jefferson, N.C.: McFarland and Co., Inc., 1992.

Congressional Quarterly, *Congress and the Nation, 1945–1964: A Review of Government and Politics in the Postwar Years*, Washington, D.C.: Congressional Quarterly Service, 1965.

Converse, Philip, and Howard Schuman, "'Silent Majorities' and the Vietnam War," *Scientific American*, June 1970, pp. 17–25.

Craig, Gordon A., and Alexander L. George, *Force and Diplomacy: Diplomatic Problems of Our Time*, 3rd Ed., New York: Oxford University Press, 1995.

Dower, John, *War Without Mercy*, New York: Pantheon, 1987.

Downs, Anthony, *An Economic Theory of Democracy*, New York: Harper and Row, 1957.

Erskine, Hazel, "Is War a Mistake?" *Public Opinion Quarterly*, Vol. 34, 1970, pp. 134–150.

Fearon, James, "Domestic Political Audiences and the Escalation of International Disputes," *American Political Science Review*, Vol. 88, No. 3, September 1994, pp. 577–592.

Gallup, George H., *The Gallup Poll: Public Opinion 1935–1971*, New York: Random House, 1972, and updated by the annual editions of *The Gallup Poll*, which are published by Scholarly Resources, Inc.

Gamson, William, and Andre Modigliani, "Knowledge and Foreign Policy Opinion," *Public Opinion Quarterly*, Vol. 30, 1966, pp. 187–199.

Garfinkle, Adam, "Aftermyths of the Antiwar Movement," *Orbis*, Fall 1995, pp. 503–516.

George, Alexander L., "Domestic Constraints on Regime Change in U.S. Foreign Policy: The Need for Policy Legitimacy," in O. R. Holsti, R. M. Siverson, and A. L. George, eds., *Change in the International System*, Boulder: Westview Press, 1980.

Goldhamer, Herbert, *The 1951 Korean Armistice Conference: A Personal Memoir*, Santa Monica, Calif.: RAND, 1994.

Hallin, Daniel, *The Uncensored War*, New York: Oxford University Press, 1986.

Hirsch, John L., and Robert B. Oakley, *Somalia and Operation Restore Hope: Reflections on Peacemaking and Peacekeeping*, Washington, D.C.: United States Institute of Peace, 1995.

Holl, Jane, *From the Streets of Washington to the Roofs of Saigon: Domestic Politics and the Termination of the Vietnam War*, Ph.D. Dissertation, Stanford University, 1989, cited in Gordon A. Craig and Alexander L. George, *Force and Diplomacy: Diplomatic Problems of Our Time*, 3rd Ed., New York: Oxford University Press, 1995, p. 244.

Hosmer, Stephen T., *Constraints on U.S. Strategy in Third World Conflict*, Santa Monica, Calif.: RAND, R-3208-AF, 1985.

Iklé, Fred C., *Every War Must End*, New York: Columbia University Press, 1991.

Inter-University Consortium for Political Research, *The 1951 Minor Election Study*, revised edition, Ann Arbor, Mich.: University of Michigan, Political Behavior Program, Survey Research Center, 1968.

Jentleson, Bruce, "The Pretty Prudent Public: Post Post-Vietnam American Opinion on the Use of Military Force," *International Quarterly*, Vol. 36, 1992, pp. 49–74.

Kagay, Michael R., "Variability Without Fault: Why Even Well-Designed Polls Can Disagree," in Thomas E. Mann and Gary R. Orren, eds., *Media Polls in American Politics*, Washington, D.C.: Brookings, 1992, pp. 95–124.

Kernell, Samuel, "Explaining Presidential Popularity: How Ad Hoc Theorizing, Misplaced Emphasis, and Insufficient Care in Measuring One's Variables Refuted Common Sense and Led the Conventional Wisdom Down the Path of Anomalies," *The American Political Science Review*, Vol. 72, No. 2, June 1978.

Key, V. O., Jr., with Milton Cummings, *The Responsible Electorate*, Cambridge, Mass.: Harvard University Press, 1966.

Klarevas, Louis, and Daniel O'Connor, *At What Cost? American Mass Public Opinion and the Use of Force Abroad*, paper presented to the Annual Meeting of the International Studies Association, 1994.

Kull, Steven, and Clay Ramsay, *U.S. Public Attitudes on Involvement in Somalia*, College Park, Md.: University of Maryland, Program on International Policy Attitudes, October 26, 1993.

Lorell, Mark, Charles Kelley, Jr., with Deborah Hensler, "Casualties, Public Opinion, and Presidential Policy During the Vietnam War," Santa Monica, Calif.: RAND, 1985.

MacKuen, Michael, "Reality, the Press, and Citizens' Political Agendas," in Charles F. Turner and Elizabeth Martin, eds., *Surveying Subjective Phenomena*, New York: Russell Sage Foundation, 1984, pp. 443–473.

McCullough, David, *Truman*, New York: Simon and Schuster, 1992.

Milstein, Jeffrey S., *Dynamics of the Vietnam War*, Columbus: Ohio State University, 1974.

Mueller, John E., "Trends in Popular Support for the Wars in Korea and Vietnam," *American Political Science Review*, June 1971, pp. 358–375.

_____, *War, Presidents, and Public Opinion*, New York: John Wiley, 1973.

_____, *Policy and Opinion in the Gulf War*, Chicago: University of Chicago Press, 1994.

National Election Studies and Inter-University Consortium for Political and Social Research (ICPSR), Institute for Social Research, *American National Election Studies, 1948–1994*, CD-ROM, Ann Arbor, Mich.: University of Michigan, May 1995.

Newman, W. Russell, *The Paradox of Mass Politics: Knowledge and Opinion in the American Electorate*, Cambridge, Mass.: Harvard University Press, 1986.

Neustadt, Richard E., "The Constraining of the President," in Aaron Wildavsky, ed., *Perspectives on the Presidency*, Boston, Mass.: Little, Brown and Company, 1975, pp. 431–446.

Office of the Secretary of Defense (Force Management and Personnel), *Manual of Military Decorations and Awards*, DoD 1348.33-M, Washington, D.C., June 1993.

O'Neill, William L., *A Democracy at War: America's Fight at Home and Abroad in World War II*, Cambridge: Harvard University Press, 1993.

Page, Benjamin I., Robert Y. Shapiro, and Glenn R. Dempsey, "What Moves Public Opinion?" *American Political Science Review*, Vol. 81, No. 1, March 1987, pp. 23–43.

Page, Benjamin I., and Robert Y. Shapiro, *The Rational Public*, Chicago: University of Chicago Press, 1992.

Petty, R. C., and J. T. Cacioppo, *Attitudes and Persuasion: Classic and Contemporary Approaches*, Dubuque, Ia.: William C. Brown Company, 1981.

_____, *Communication and Persuasion: Central and Peripheral Routes to Persuasion*, New York: Springer-Verlag, 1986.

Richman, Alvin, "When Should We Be Prepared to Fight?" *The Public Perspective*, April/May 1995, pp. 44–49.

Rielly, John E., ed., *American Public Opinion and U.S. Foreign Policy 1995*, Chicago: The Chicago Council on Foreign Relations, 1995.

Russett, Bruce, and Miroslav Nincic, "American Opinion on the Use of Military Force Abroad," *Political Science Quarterly*, Vol. 91, No. 3, 1976, pp. 411–431.

Schwarz, Benjamin J., *Casualties, Public Opinion and U.S. Military Intervention: Implications for U.S. Regional Deterrence Strategies*, Santa Monica, Calif.: RAND, 1994. OUT OF PRINT. Superseded by this document, MR-726-RC.

Smoke, Richard, "On the Importance of Policy Legitimacy," *Political Psychology*, Vol. 15, No. 1, 1994, pp. 97–110.

"Statement by Press Secretary Fitzwater on United States Military Action in Panama, December 21, 1989," in *Administration of George Bush, 1989*, Washington, D.C.: Government Printing Office, 1990, pp. 1726–1727.

"Statement by Press Secretary Fitzwater on Additional Humanitarian Aid for Somalia, August 14, 1992," in *Administration of George Bush, 1992*, Washington, D.C.: Government Printing Office, 1993, p. 1360.

Voth, Alden, "Vietnam: Studying a Major Controversy," *Journal of Conflict Resolution*, Vol. 11, No. 4, pp. 431–443.

Yankelovich, Daniel, *Coming to Public Judgment: Making Democracy Work in a Complex World*, Syracuse, N.Y.: Syracuse University Press, 1991.

Zaller, John R., *The Role of Elites in Shaping Public Opinion*, unpublished doctoral dissertation, University of California, Berkeley, 1984.

_____, "The Diffusion of Political Attitudes," *Journal of Personality and Social Psychology*, Vol. 53, 1987, pp. 821–833.

_____, *The Nature and Origins of Mass Opinion*, Cambridge: Cambridge University Press, 1992.

_____, "The Converse-McGuire Model of Attitude Change and the Gulf War Opinion Rally," *Political Communication*, Vol. 10, 1993, pp. 369–388.

Zelman, Walter A., *Senate Dissent and the Vietnam War, 1964–1968*, doctoral dissertation, Ann Arbor, Mich.: University Microfilms, 1971.